Assessing Innovation

We cannot manage and control what we cannot measure and assess. Poor assessment results have been cited as a primary reason for project failures, in terms of cost, schedule, and quality. This book introduces metrics, rubrics, and standards pertinent to assessing innovation.

Assessing Innovation: Metrics, Rubrics, and Standards provides a new view for the embracing of innovation and establishing a quantitative basis for determining innovation levels. It bridges innovation with practice and presents a systems view as it incorporates the human element and discusses the different roles carried out. This book offers standards that will guide readers as they tackle sustaining innovation and leverages Badiru's Umbrella Model in the process.

The inclusion of methodologies suitable for determining where and when innovation is happening, and to what extent it is currently being carried out, make this a unique book, along with being the only book that addresses innovation metrics, rubrics, and standards in an integrated fashion. Seen as a way to help advance the diverse pursuit of innovation, this book is an ideal read for those in engineering, business, industry, academia, government, and the military.

T0351022

Systems Innovation Book Series

Series Editor: Adedeji Badiru

Systems Innovation refers to all aspects of developing and deploying new technology, methodology, techniques, and best practices in advancing industrial production and economic development. This entails such topics as product design and development, entrepreneurship, global trade, environmental consciousness, operations and logistics, introduction and management of technology, collaborative system design, and product commercialization. Industrial innovation suggests breaking away from the traditional approaches to industrial production. It encourages the marriage of systems science, management principles, and technology implementation. Particular focus will be the impact of modern technology on industrial development and industrialization approaches, particularly for developing economics. The series will also cover how emerging technologies and entrepreneurship are essential for economic development and society advancement.

Global Supply Chain
Using Systems Engineering Strategies to Respond to Disruptions

Adedeji B. Badiru

Systems Engineering Using the DEJI Systems Model®
Evaluation, Justification, Integration with Case Studies and Applications

Adedeji B. Badiru

Handbook of Scholarly Publications from the Air Force Institute of Technology (AFIT), Volume 1, 2000-2020
Edited by Adedeji B. Badiru, Frank Ciarallo, and Eric Mbonimpa

Project Management for Scholarly Researchers
Systems, Innovation, and Technologies

Adedeji B. Badiru

Industrial Engineering in Systems Design
Guidelines, Practical Examples, Tools, and Techniques

Brian Peacock and Adedeji B. Badiru

Leadership Matters
An Industrial Engineering Framework for Developing and Sustaining Industry

Adedeji B. Badiru and Melinda Tourangeau

Systems Engineering
Influencing Our Planet and Reengineering Our Actions

Adedeji B. Badiru

Total Productive Maintenance, Second Edition
Strategies and Implementation Guide

Tina Agustiady and Elizabeth A. Cudney

Assessing Innovation
Metrics, Rubrics and Standards

Adedeji B. Badiru and Melinda Tourangeau

Assessing Innovation

Metrics, Rubrics, and Standards

Adedeji B. Badiru

Melinda L. Tourangeau

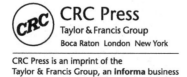

CRC Press
Taylor & Francis Group
Boca Raton London New York

CRC Press is an imprint of the
Taylor & Francis Group, an **informa** business

Designed cover image: Adedeji B. Badiru

First edition published 2024
by CRC Press
2385 NW Executive Center Drive, Suite 320, Boca Raton FL 33431

and by CRC Press
4 Park Square, Milton Park, Abingdon, Oxon, OX14 4RN

CRC Press is an imprint of Taylor & Francis Group, LLC
© 2024 Adedeji B. Badiru and Melinda L. Tourangeau

ISBN: 978-1-032-51418-5 (hbk)
ISBN: 978-1-032-51702-5 (pbk)
ISBN: 978-1-003-40354-8 (ebk)

DOI: 12.01/9781003403548

Typeset in Times New Roman
by Deanta Global Publishing Services, Chennai, India

Dedicated to all the innovators, past and present, who have helped shape the world we now live in, play in, and enjoy.

Contents

Acknowledgements

We acknowledge the support and encouragement of our family members (immediate and extended), friends, and professional colleagues for their motivational influence in embarking on this literary journey, addressing innovation from a measurement and assessment standpoint. We appreciate the support of our CRC Press Executive Editor, Cindy Carelli, for welcoming and embracing our brazen thoughts of delving into the metrics and rubrics of innovation, a topic not hitherto addressed in the literature. Without her reception of our idea, this book would not have happened.

Foreword

Innovation is the lifeblood of progress and the driving force behind advancements in every industry and field. It is the spark that ignites creativity, propels us forward, and shapes the world we live in. The often-asked questions are:

- How do we measure innovation?
- How do we quantify its impact?
- How do we harness its power to drive growth and success?

In *Assessing Innovation: Metrics, Rubrics, and Standards*, authors Adedeji Badiru, PhD and Melinda Tourangeau dive deep into the world of innovation assessment, providing a comprehensive guide for understanding and evaluating the effectiveness of innovative efforts. Through a blend of theoretical foundations, practical examples, and real-world case studies, the authors explore the various dimensions of innovation measurement, offering valuable insights and tools for those leading the way to innovation.

Authored by Adedeji Badiru, PhD, and Melinda Tourangeau, this technical reference book provides a comprehensive guide to assessing and quantifying innovation. As two esteemed experts in their respective fields, the authors have combined their knowledge and experience to create a valuable resource for students, professors, researchers, consultants, entrepreneurs, policymakers, marketers, and other professionals engaged in the realm of innovation.

The book's table of contents reveals a comprehensive exploration of the subject, covering everything from the fundamentals of measurements and rubrics to the economic analysis of innovation. Each chapter is meticulously crafted, offering an awareness and understanding of the intricacies of innovation assessment. From defining metrics and standards to utilizing statistical engineering approaches, the authors leave no stone unturned, equipping readers with the tools they need to evaluate and enhance their innovative endeavors.

What sets this book apart is its practicality. While other texts may delve into the theoretical aspects of innovation measurement, *Assessing Innovation: Metrics, Rubrics, and Standards* takes a hands-on approach. The authors provide practical examples and case studies that bring the concepts to life, enabling readers to apply the knowledge gained to their unique contexts. This emphasis on real-world application differentiates this book, thus making it a valuable resource for those involved in delivering innovation.

As the foreword authors, we have had the privilege of witnessing the authors' dedication and expertise firsthand. Their commitment to advancing the field of innovation assessment is evident on every page of this book. It is our firm belief that *Assessing Innovation: Metrics, Rubrics, and Standards* will serve as an invaluable resource for anyone seeking to understand, evaluate, and enhance their innovative efforts.

We wholeheartedly recommend this book to students, professors, researchers, consultants, entrepreneurs, policymakers, marketers, and other professionals engaged with innovation. Whether you are just beginning your journey in the field or are a seasoned expert, the insights and tools provided within these pages will undoubtedly prove indispensable.

We would like to commend Adedeji Badiru, PhD, and Melinda Tourangeau for their exceptional work in producing a technical reference book that will undoubtedly shape the future of innovation assessment. Their dedication to the advancement of the field is commendable, and we are confident that *Assessing Innovation* will become a go-to resource for years to come.

Dana W. Clarke, *Sr., President/CEO Applied Innovation Alliance*
Donald M. Reimer, *President, The Small Business Strategy Group*

Preface

Innovation has been the bastion of societal advancement for generations. Innovation, in one form or another, is the foundation for the national development we enjoy today. However, if not addressed as a topical focus in today's socially and politically charged environment, we may begin to see an erosion of the basic tenets of innovation. We must explicitly rededicate ourselves to "innovation" as a societal focus. This book, which is designed to serve as a textbook as well as a reference book, provides the fundamental pathways to instituting and managing innovation with a world-wide perspective. The recent emergence of innovation parks, centers, hubs, institutes, partnerships, consortia, and so on attests to the growing need to pay more attention to innovation. Almost all of the existing books in the market on innovation are patterned as reference books. There is a lack of books offering a rigorous and integrated presentation of the science, technology, engineering, and mathematics of innovation superimposed on managerial processes. The technical and managerial expertise and experience of the authors provide the basis for writing the proposed book to address the existing void. Innovation in the military is transformed to business and industry for the advancement of the society through rational technology transfer paths. Innovation in academia, government, and human–computer interactions leads to much-desired consumer products. In this regard, the contents of this book are of interest to all and sundry throughout the broad spectrum of embracing innovation.

Adedeji B. Badiru
Melinda L. Tourangeau

About the Authors

Adedeji B. Badiru is a Professor of Systems Engineering at the Air Force Institute of Technology and a registered professional engineer. He is also a fellow of the Institute of Industrial Engineers and Fellow of the Nigerian Academy of Engineering. Dr. Badiru has a BS degree in Industrial Engineering, an MS in Mathematics, an MS in Industrial Engineering from Tennessee University, and a PhD in Industrial Engineering from the University of Central Florida. He is the author of several books and technical journal articles and has received several awards and recognitions for his accomplishments. He is also a series editor for CRC Press/Taylor and Francis.

Melinda L. Tourangeau is the Executive Director of the RVJ Institute, a 501 c (3) center of excellence and research institute dedicated exclusively to excellence in the electromagnetic environment. She possesses advanced degrees in Electrical Engineering and Business Administration and is currently pursuing a PhD in Education with an emphasis on organizational systems. Ms. Tourangeau is considered a subject matter expert in Electromagnetic Warfare and Electromagnetic Spectrum Operations. She has authored numerous reports for the Department of Defense and US Congress and given presentations to audiences in Europe, Hawaii, Canada, and the United States. Her background emphasizes electro-optics, lasers, and semiconductor physics, as well as organizational and leadership systems. Her career includes serving as a Department of Defense program manager for critical Electromagnetic Warfare programs and serving in the U.S. Air Force.

1 Fundamentals of Measurements and Rubrics

THE PERVASIVENESS OF INNOVATION

Ready! Get set! Go! The innovation race is on, and it is gaining speed. Innovation is all around us, in verbiage, promulgations, and actions, making it difficult to discern whether innovation is really taking place among the barrage of proclamations about it. Everywhere we turn these days, everyone is proclaiming themselves to be doing it and/or pursuing it. All manner of progress is showcased with innovation, such as "scaling up innovation," "exercising innovation," and "more/significant/faster innovation," just to name a few. There is a plethora of innovation symposia, fora, conventions, conferences, seminars, and meetings all over the world. Likewise, innovation *corridors*, *centers*, *institutes*, and *platforms* have emerged in recent years. Even occupational job series are emerging, with titles such as "Innovation Coordinator," "Innovation Facilitator," "Innovation Accelerator," etc. The website www.etymonline.com/word/innovation presents an interesting view of innovation, what it is and what it portends (good and bad). The mere use of the word has exploded exponentially over the last 70 years (see Figure 1.1). It appears that innovation has emerged as a free-for-all pursuit. Simply put, innovation's etymology implies "to change; to renew. A meaning from the 1540s indicates innovation as "a novel change, experimental variation, new thing introduced in an established arrangement." Innovation has an element of being life-changing; in other words, a tremendous contribution to the quality of life, such that things will never be the same. Just that thought alone conjures up a great deal of excitement. Engineers, scientists, inventors, analysts, and leaders reading this book will probably feel a rush of energy from contemplating the potential of their endeavors when innovation makes its way into the process. This is probably best summed up by the following quote:

> While necessity is the mother of invention, creativity is the father of innovation.
>
> Melinda L. Tourangeau, 2023

Our willingness to create, to test the art of the possible, to take risks, to experiment, and to take pleasure in our achievements fans the flames of creativity to spark innovation. An invention remains just a great idea until it is innovated. When innovation

DOI: 10.1201/9781003403548-1

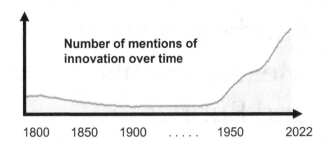

FIGURE 1.1   Conceptual Plot of Number of Mentions of Innovation over Time

arrives, the world is forever changed. That will be just one of the metrics proposed in this book. Are you ready to change the world? Keep reading ….

In all of this scrambling about innovation, one thing that is consistently missing is the measurement and affirmation of innovation. Is it sufficient to merely incarnate the word "innovation" and expect the world to believe? Of course not. Anchoring innovation in metrics, rubrics, and standards can benefit both progenitors and society. Therein lies the genesis of this book. Innovation conjures the presence of a new method, idea, product, etc. Innovation metrics are therefore essential. How do we know when we have arrived at or achieved innovation, granted that innovation is essential for the continuing success of any organization?

IMPORTANCE OF MEASUREMENTS IN INNOVATION

After all the preceding discussions, the questions that come to mind include the following:

- How do we measure innovation on a scale?
- How do we affirm each claim of innovation?
- Does innovation always connote the same thing across every spectrum?

If something has not been (or cannot be) measured and affirmed, its authenticity and value will always be debatable. For this reason, this book intends to set out a solid framework for measuring and "grading" innovation. It offers a comprehensive description of measurements that exist today, with the intention for the reader to gain a full understanding of the importance of measurements (Badiru and Racz, 2016) and their implications for bolstering innovation in business, industry, government, and the military. Innovation is pervasive and multi-definitional. Let us consider the following:

> Innovation is seeing what everybody has seen and thinking what nobody has thought.
>
> Dr. Albert Szent-Gyorgyi

This quote is akin to building a new mousetrap after all the conventional mice have been annihilated. Throughout history, humans have striven to come up with better tools, techniques, and instruments for measurement. From the very ancient times to the present fast-paced society, our search for more precise, more convenient, and more accessible measuring devices has led to new developments over the years. The following quotes confirm the importance and efficacy of measurements in our lives. The Appendix presents the most common measurement conversion factors.

"Measure twice, cut once" – English Proverb
"Where there is no Standard there can be no Kaizen (improvement)" – Taiichi Ohno
"Where there is no measurement, there can be no standard" – Adedeji Badiru
"Measurement mitigates mess-ups" – Adedeji Badiru

Measurement pervades everything we do. This applies to technical, management, and social activities and requirements. Even in innocuous situations, such as human leisure, the importance of measurement comes to the surface. How much, how far, how good, how fast, and how long are typical conveyances of some sort of measurement. Consider a possible newspaper classified advertisement that provides the following measurement disclaimer:

The acceptable units of measure for all firewood advertisement are cord or fraction of a cord. The units of measurements for a cord are 4″ by 4″ by 8″. The terms face cord, rack, pile, rick, truckload, or similar terms are not acceptable.

A face cord is a measurement of wood in the USA and Canada. It is a standardized unit of volume that has been defined by law, which means it can be bought or sold with confidence. On the other hand, a cord of wood is one standard length (4 feet) of firewood. Who would have thought that firewood had such a serious measurement constraint and guideline? Social, psychological, physical, economic, cognitive, and metabolic attributes as well as other human characteristics are all amenable to measurement systems, just like the mechanical devices around us.

Every item that we recognize as innovation is subject to some sort of measurement of the attributes of the item. This could be its length, weight, height, color, volume, time dimension, profit potential, and so on. Consequently, to have confidence in whatever we judged to be an innovation, it is essential to think in terms of the pertinent measurements of the characteristics of that item. So, we need to understand what a measurement represents, whether it is on a quantitative scale or a qualitative scale. Not only must a measurement be made; it must also be articulated via communication so that the audience can fully appreciate the item to which the measurement pertains. In this regard, we offer the following inspirational quote:

While necessity is the mother of invention, communication is the spark for innovation.
Adedeji Badiru, 2023

The importance of communication for innovation, from a systems perspective, will be further elucidated in a later section on a systems approach to innovation using the DEJI Systems Model in Chapter 4.

ANALYSIS OF A MEASUREMENT

It is well understood that we cannot manage anything if we cannot measure it. All elements involved in our day-to-day decision making involve some form of measurement. Measuring an attribute of a system and then analyzing it against some standard, some best practice, or some benchmark empowers a decision-maker to take appropriate and timely actions.

Fundamentally, measurement is the act, or the result, of a quantitative comparison between a predefined standard and an unknown magnitude. Beckwith and Buck (1965), Shillito and De Marle (1992), Morris (1997), Badiru et al. (2012), and Badiru and Kovach (2012) all address concepts, tools, and techniques of measurement systems. If the result is to be generally meaningful, two requirements must be met in the act of measurement:

1. The standard that is used for comparison must be accurately known and commonly accepted.
2. The procedure and instrument employed for obtaining the comparison must be provable and repeatable.

The first requirement is that there is an accepted standard of comparison. A weight cannot simply be heavy. It can only be proportionately heavy to something else, namely, the standard. A comparison must be made, and unless it is made relative to something generally recognized as standard, the measurement can only have limited meaning. This holds for any quantitative measurement we may wish to make. In general, the comparison is one of magnitude, and a numerical result is presupposed. The quantity in question may be twice as large as the standard, or 1.4 times as large, or in some other ratio, but a numerical comparison must be made for it to be meaningful. The typical characteristics of a measurement process include the following:

- Precision
- Accuracy
- Correlation
- Stability
- Linearity
- Type of Data

THE PAST AND PRESENT OF MEASUREMENT

Weights and measures may be ranked among the necessaries of life to every individual of human society. They enter into the economical arrangements and daily concerns of every family. They are necessary to every occupation of human industry; to the distribution and security of every species of property; to every transaction of trade and commerce; to the labors of the husbandman; to the ingenuity of the artificer; to the studies of the philosopher; to the researches of the antiquarian; to the navigation of the mariner, and the marches of the soldier; to all the exchanges of peace, and all the operations of war. The knowledge of them, as in established use, is among the first elements of education, and is often learned by those who learn nothing else, not even to read and write. This knowledge is riveted in the memory by the habitual application of it to the employments of men throughout life.

– John Quincy Adams, Report to the Congress, 1821

The historical accounts of measurements presented in this section are based mostly on USA National Institute of Standards and Technology (NIST, 1974), HistoryWorld (2014), and Ta Neter Foundation (TANF, 2014). Weights and linear measures were among the earliest tools invented by man. Primitive societies needed rudimentary measures for many tasks, such as house and road construction and commerce of raw materials. Man, in early years, used parts of the human body (usually the ruling class's own) and the natural surroundings to devise measuring standards. Early Babylonian and Egyptian records and the Bible indicate that length was first measured with the foot, forearm, hand, or finger. Time was measured by the periods of the sun, moon, and other heavenly bodies. When it was necessary to compare the capacities of containers such as gourds of clay or metal vessels, they were filled with plant seeds that were then counted to measure volumes. With the development of scales as a means for weighing, seeds and stones served as standards. For instance, the "carat," still used as a mass unit for gems, is derived from the carob seed.

As societies evolved, measurements became more complex. The invention of numbering systems and the science of mathematics made it possible to create whole systems of measurement units suitable for trade and commerce, land division, taxation, and scientific research. For these more sophisticated uses, it was necessary not only to weigh and measure more complex items, but also to do so accurately and repeatedly at different locations. Because of limited international exchange of goods and communication of ideas in ancient times, different measuring systems evolved for the same measures and became established in different parts of the world. In this same time period, even different parts of the same country might use different measuring systems for the same purpose. Historical records indicate that early measurement systems evolved locally in Africa to take advantage of the African natural environment. For example, common early units of measure in some parts of Africa relied on standardizations based on the size and weight of cocoa beans.

METRICS: A COLLECTION OF MEASUREMENTS

The title of this book is *Assessing* Innovation – *Metrics*, Rubrics, and Standards. To assess means to measure, and the way a field of interest or study collates those measurements is through a taxonomized collection of well-defined and relevant measurements. When those measurements are adopted into a model or methodology to ensure quality, gauge outcomes, or meet performance standards of the field of study, they become *metrics*. This may seem parochial, but it is worth mentioning if this book is going to meet its purpose. This book is in itself innovative, in that it will invite the world to view innovation in a whole new way. No longer will "claims to be innovative" be the marker or the bar. Very soon, and as a result of the metrics being proposed in this book, companies will need to "measure up" by examining, evaluating, and measuring their actions when it comes to innovation, and using those results to prove they have innovated, or they are innovative. As with nearly everything occurring in our digital era now, these measurements will be backed up by reliable data analysis. The latter portion of this chapter expounds on the considerations of data analysis. Its importance should not be overlooked, and in fact, data analysis should be made paramount when it comes to taking on claiming innovation in the 21st century.

THE ENGLISH SYSTEM OF MEASUREMENT

The measurement system commonly used in the USA today is nearly the same as that brought by the American colony settlers from England. These measures had their origins in a variety of cultures, including Babylonian, Egyptian, Roman, Anglo-Saxon, and Nordic French. The ancient "digit," "palm," "span," and "cubic" units of length slowly lost preference to the length units "inch," "foot," and "yard." Roman contributions include the use of 12 as a base number and the words from which we derive many of the modern names of measurement units. For example, the 12 divisions of the Roman "pes," or foot, were called "unciae." The "foot," as a unit of measuring length, is divided into 12 inches. The common words "inch" and "ounce" are both derived from the same Latin word, "uncia," meaning one-twelfth. The "yard" as a measure of length can be traced back to early Saxon kings. They wore a sash or girdle around the waist that could be removed and used as a convenient measuring device. Thus, the word "yard" comes from the Saxon word "gird," which represents the circumference of a person's waist, preferably a "standard person," such as a king.

The titles for measurements aside, the evolution and standardization of measurement units often had interesting origins as well. For example, it was recorded that King Henry I decreed that a yard should be the distance from the tip of his nose to the end of his outstretched thumb. The length of a furlong (or furrow-long) was established by early Tudor rulers as 220 of these same King Henry I yards. This led Queen Elizabeth I to declare in the 16th century that the traditional Roman mile of 5,000 feet would be replaced by one of 5,280 feet, making the mile exactly 8 furlongs and providing a convenient relationship between the furlong and the mile. To this day, there are 5,280 feet in 1 mile, which is 1,760 yards. Thus, through royal edicts, England by the 18th century had achieved a greater degree of standardization than other European countries. The promulgation of English units was improved by the discovery of coal in England and the subsequent transmission of coal to many parts of the world, bringing the dawn of the First Industrial Revolution. The English units were well suited to burgeoning commerce and trade because they had been developed and refined to meet commercial needs. Through English colonization and its dominance of world commerce during the 17th, 18th, and 19th centuries, the English system of measurement units became established in many parts of the world, including the American colonies. The early 13 American colonies, however, had undesirable differences with respect to measurement standards for commerce. The need for greater uniformity led to clauses in the Articles of Confederation (ratified by the original colonies in 1781) and the Constitution of the United States (ratified in 1788) that gave Congress the power to fix uniform standards for weights and measures across the colonies. Today, standards provided by the NIST ensure uniformity of measurement units throughout the country.

THE METRIC SYSTEM OF MEASUREMENT

The need for a single worldwide coordinated measurement system was recognized over 300 years ago. In 1670, Gabriel Mouton, Vicar of St. Paul's Church in Lyons, France, and an astronomer, proposed a comprehensive decimal measurement system based on the length of 1 minute of arc of a great circle of the Earth. Mouton also proposed the swing length of a pendulum with a frequency of one beat per second as the unit of length. A pendulum with this beat would have been fairly easily reproducible, thus facilitating the widespread distribution of uniform standards.

In 1790, in the midst of the French Revolution, the National Assembly of France requested the French Academy of Sciences to "deduce an invariable standard for all the measures and all the weights." The Commission appointed by the Academy created a system that was, all at once, simple and scientific. The unit of length was to be a portion of the Earth's circumference. Measures for capacity (volume) and mass were to be derived from the unit of length, thus relating the basic units of the system to each other and to nature. Furthermore, larger and smaller multiples of each unit were to be created by multiplying or dividing the basic units by 10 and powers of 10. This feature provided a great convenience to users of the system by eliminating the need for such calculations as dividing by 16 (to convert ounces to pounds) or by 12 (to convert inches to feet). Similar calculations in the metric system could be performed simply by shifting the decimal point. Thus, the metric system is a "base-10" or "decimal" system.

The Commission assigned the name "metre" (i.e., meter in English) to the unit of length. This name was derived from the Greek word *metron*, meaning "a measure." The physical standard representing the meter was to be constructed so that it would equal one ten-millionth of the distance from the North Pole to the equator along the meridian running near Dunkirk in France and Barcelona in Spain. The initial metric unit of mass, the "gram," was defined as the mass of 1 cubic centimeter (a cube that is 0.01 meter on each side) of water at its temperature of maximum density. The cubic decimeter (a cube 0.1 meter on each side) was chosen as the unit for capacity. The fluid volume measurement for the cubic decimeter was given the name "liter." Although the metric system was not accepted with much enthusiasm at first, adoption by other nations occurred steadily after France made its use compulsory in

1840. The standardized structure and decimal features of the metric system made it well suited for scientific and engineering work. Consequently, it is not surprising that the rapid spread of the system coincided with an age of rapid technological development. In the USA, by Act of Congress in 1866, it became "lawful throughout the United States of America to employ the weights and measures of the metric system in all contracts, dealings or court proceedings." However, the USA has remained a hold-out with respect to a widespread adoption of the metric system. Previous attempts to standardize to the metric system in the USA have failed. Even today, in some localities of the USA, both English and metric systems are used side by side.

As a good illustration of dual use (and subsequent chaos) of measuring systems, a widespread news report in late September 1999 exposed how the National Aeronautics and Space Administration (NASA) lost a $125 million Mars orbiter in a crash onto the surface of Mars because a Lockheed Martin engineering team used the English units of measure while the agency's team used the more conventional metric system for a key operation of the spacecraft. This unit mismatch prevented mutually understood navigation information from being transferred between the Mars Climate Orbiter spacecraft team at Lockheed Martin in Denver and the flight team at NASA's Jet Propulsion Laboratory in Pasadena, California. As this case demonstrates, even high-stakes scientific endeavors are not spared from the havoc that can emerge from not standardizing which units of measure to use.

Getting back to history, the late 1860s saw the need for even better metric standards to keep pace with scientific advances. In 1875, an international agreement, known as the Meter Convention, set up well-defined metric standards for length and mass and established permanent mechanisms to recommend and adopt further refinements in the metric system. This agreement, commonly called the "Treaty of the Meter" in the USA, was signed by 17 countries, including the USA. As a result of the Treaty, metric standards were constructed and distributed to each nation that ratified the Convention. Since 1893, the internationally adopted metric standards have served as the fundamental measurement standards of the USA, at least in theory if not in practice.

By 1900, a total of 35 nations, including the major nations of continental Europe and most of South America, had officially accepted the metric system. In 1960, the General Conference on Weights and Measures, the diplomatic organization made up of the signatory nations to the Meter Convention, adopted an extensive revision and simplification of the system. The following seven units were adopted as the base units for the metric system:

1. Meter (for length)
2. Kilogram (for mass)
3. Second (for time)
4. Ampere (for electric current)
5. Kelvin (for thermodynamic temperature)
6. Mole (for amount of substance)
7. Candela (for luminous intensity)

ELEMENTS OF THE SI SYSTEM

Based on the general standardization described in the previous section, the name Système International d'Unités (International System of Units), with the international abbreviation SI, was adopted for the modern metric system. Throughout the world, measurement science research and development continues to develop more precise and easily reproducible ways of defining measurement units. The working organizations of the General Conference on Weights and Measures coordinate the exchange of information about the use and refinement of the metric system and make recommendations concerning improvements to the system and its related standards. Our daily lives are mostly ruled or governed by the measurements of length, weight, volume, and time. These are briefly described in the following sections.

LENGTH MEASUREMENT

The measurement of distance, signified as length, is the most ubiquitous measurement in our world. The units of length represent how we conduct everyday activities and transactions. The two basic units of length measurement are the British units (inch, foot, yard, mile) and the metric system (meters, kilometers). In its origin, the inch is anatomically a thumb. The foot logically references the human foot. The yard relates closely to a human pace, but also derives from two cubits (the measure of the forearm). The mile originates from the Roman *mille passus*, which means a thousand paces. The ancient Romans defined a pace as two steps. So, approximately, a human takes two paces within 1 yard. The average human walking speed is about 5 kilometers per hour (km/h), or about 3.1 miles per hour (mph).

For the complex measuring problems of civilization – surveying land to register property rights, or selling a commodity by length – a more precise unit is required. The solution is a rod or bar, of an exact length, kept in a central public place. From this "standard," other identical rods can be copied and distributed through the community. In Egypt and Mesopotamia, these standards were kept in temples. The basic unit of length in both civilizations was the cubit, based on a forearm measured from the elbow to the tip of the middle finger. When a length such as this was standardized, it was usually the king's dimension that was first taken as the norm.

WEIGHT MEASUREMENT

For measurements of weight, the human body does not provide convenient approximations as we have for length. For comparative purposes, it turns out that grains of wheat are reasonably standard in size. Weight can be expressed with some degree of accuracy in terms of a number of grains of wheat. The use of grains to convey weight is still used today in the measurement of precious metals, such as gold, by jewelers. As with measurements of length, a block of metal can be kept in the temples as an official standard for a given number of grains. Copies of this can be cast and weighed in the balance for perfect accuracy. However, imperfect human integrity in using scales made it necessary to have an inspectorate of weights and measures for practical adjudication of measurements in the olden days.

WEIGHT OR MASS?

It is prudent to remind ourselves and distinguish between the measure of "weight" and the measure of "mass." Weight is a unit of force; an item's "weight" on the surface of the earth is determined by the force of gravity being exerted on it. This, of course, means that the same item's weight on the moon will be different from its weight on the earth. The earth exerts a downward force on all objects resting on or moving on the earth's surface (or even near it), and it matters not whether the objects are in motion or still; the weight is the same.

Weight can be calculated by multiplying the mass (m) of an object by gravity, g, or

$$F_g = mg = W$$

The metric units of weight are kg*m/s^2.

Mass is the measure of the *inertia* of an object. Inertia is a property of a body or mass such that the object will continue in its current state until it is disrupted by an external force. The units for inertia are kg*m^2, where kg is kilograms, the *mass* of the object in metric units, and m^2 is "meters squared," or the square of the distance from the axis of rotation to the particles that make up the object. Simply put, the more mass an object has, the greater its inertia, and the harder it is to move. To summarize, weight is a unit of *force*, and mass is a unit of *inertia*.

VOLUME MEASUREMENT

From the ancient time of commerce and trading to the present day, a reliable standard of volume is one of the hardest to accomplish. However, we improvise by using items from nature and art. Items such as animal skins, baskets, sacks, or pottery jars can be made to approximately consistent sizes, such that they are sufficient for measurements in ancient measurement transactions. Where the exact amount of any commodity needs to be known, weight is the measure more likely to be used instead of volume.

TIME MEASUREMENT

Time is a central aspect of human life. Throughout human history, time has been appreciated in very precise terms. Due to the celestial precision of day and night, the day and the week are easily recognized and recorded. However, an accurate calendar for the year is more complicated to achieve universally. The morning time before midday (forenoon) is easily distinguishable from the time after midday (afternoon), provided the sun is shining, and the position of the sun in the landscape can reveal roughly how much of the day has passed. By contrast, the smaller units of time, such as hours, minutes, and seconds, were initially (in ancient times) unmeasurable and unneeded; unneeded because ancient man had big blocks of time to accomplish whatever needed to be done. Micro-allocation of time was, thus, not essential. However, in our modern society, the minute (tiny) time measurement of seconds and minutes is essential. The co-editor's poem reprinted here conveys the modern appreciation of the passage of time:

The Flight of Time
What is the speed and direction of Time?
Time flies; but it has no wings.
Time goes fast; but it has no speed.
Where has time gone? But it has no destination.
Time goes here and there; but it has no direction.
Time has no embodiment. It neither flies, walks, nor goes anywhere.
Yet, the passage of time is constant.

Adedeji Badiru, 2006

MEASUREMENTS IN THE ERA OF SUNDIAL AND WATER CLOCK

The sundial and water clock originated in the second millennium BC. The movement of the sun through the sky makes possible a simple estimate of time from the length and position of a shadow cast by a vertical stick. If marks were made where the sun's shadow fell, the time of day could be recorded in a consistent manner. The result was the sundial. An Egyptian example survives from about 800 BC, but records indicate that the principle was familiar to astronomers of earlier times. In practice, it is difficult to measure time precisely on a sundial because the sun's path through the sky changes with the seasons. Early attempts at precision in timekeeping relied on a different principle known as the water clock. The water clock, known from a Greek word as the *clepsydra*, attempted to measure time by the amount of water that dripped from a tank. This would be a reliable form of clock if the flow of water could be perfectly controlled. In practice, at that time, it could not. The hourglass, using sand on the same principle, had an even longer history and utility. It was a standard feature used on 18th-century pulpits in Britain to ensure a sermon of standard and sufficient duration.

ORIGIN OF THE HOUR

The hour as a unit of time measurement originated in the 14th century. Until the arrival of clockwork in the 14th century AD, an hour was a variable concept. It is a practical division of the day into 12 segments (12 being the most convenient number for dividing into fractions, since it is divisible by 2, 3, and 4). For the same reason, 60, divisible by 2, 3, 4, and 5, has been a larger framework of measurement ever since the Babylonian times. The traditional concept of the hour, as one-twelfth of the time between dawn and dusk, is useful in terms of everyday timekeeping. Approximate appointments are easily made at times that are easily sensed. Noon is always the sixth hour. Half way through the afternoon is the ninth hour. This is famous as the time of the death of Jesus on the Cross. The trouble with the traditional hour is that it differs in length from day to day. Furthermore, a daytime hour is different from

one in the night (also divided into 12 equal hours). A clock cannot reflect this variation, but it can offer something more useful. It can provide every day something that
occurs naturally only twice a year, at the spring and autumn equinoxes, when the
12 hours of day and the 12 hours of night are the same length. In the 14th century,
coinciding with the first practical clocks, the meaning of an hour gradually changed.
It became a specific amount of time, one-twenty-fourth of a full solar cycle from
dawn to dawn. Today, the day is recognized as 24 hours, although it still features on
clock faces as two 12s.

MINUTES AND SECONDS IN THE 14TH TO 16TH CENTURY

Minutes and seconds, as we know them today, originated from the 14th to the
16th centuries. Even the first clocks could measure periods less than an hour, but
soon, striking the quarter-hours seemed insufficient. With the arrival of dials for
the faces of clocks, in the 14th century, something like a minute was required. The
Middle Ages inherited a scale of scientific measurement based on 60 from Babylon.
In Medieval Latin, the unit of one-sixtieth is *pars minuta prima* ("first very small
part"), and a sixtieth of that is *pars minute secunda* ("second very small part"). Thus,
on a principle that is 3,000 years old, minutes and seconds find their way into our
modern time. Minutes were mentioned from the 14th century, but clocks were not
precise enough for "seconds" of time to be needed until two centuries later.

MEASUREMENT BY HERO'S DIOPTRA

Hero's Dioptra was from the first century AD. One of the surviving books of Hero
of Alexandria, entitled *On the Dioptra*, describes a sophisticated technique which he
developed for surveying land. Plotting the relative position of features in a landscape
(essential for any accurate map) is a more complex task than simply measuring distances. It was necessary to discover accurate angles in both the horizontal and vertical planes. To make this possible, a surveying instrument had to somehow maintain
both planes consistently in different places, so as to take readings of the deviation
in each plane between one location and another. This is what Hero achieved with
the instrument mentioned in his title, the *dioptra*, which approximately means the
"spyhole," through which the surveyor looks when pinpointing the target in order
to read the angles. For his device, Hero adapted an instrument long used by Greek
astronomers (e.g., Hipparchus) for measuring the angle of stars in the sky. In his day,
Hero achieved his device without the convenience of two modern inventions, the
compass and the telescope.

MEASUREMENTS OF BAROMETER AND ATMOSPHERIC PRESSURE

The barometer and the measurement of atmospheric pressure originated in 1643–
1646. Like many significant discoveries, the principle of the barometer was observed
by accident. Evangelista Torricelli, assistant to Galileo at the end of his life, was
interested in why it is more difficult to pump water from a well in which the water

lies far below ground level. He suspected that the reason might be the weight of the extra column of air above the water, and he devised a way to test this theory. He filled a glass tube with mercury, an element that is 13.6 times more dense than water, thus saving Torricelli from needing a tube greater than 30 feet high. Submerging it in a bath of mercury and raising the sealed end to a vertical position, he found that the mercury slipped a little way down the tube. He reasoned that the weight of air on the mercury in the bath was supporting the weight of the column of mercury in the tube. If this was true, then the space in the glass tube above the mercury column must be a vacuum. This rushed him into controversy with traditional scientists of the day, who believed nature abhorred a vacuum. But it also encouraged von Guericke, in the next decade, to develop the vacuum pump. The concept of varying atmospheric pressure occurred to Torricelli when in 1643, he noticed that the height of his column of mercury sometimes varied slightly from its normal level, which was 760 mm above the mercury level in the bath. Observation suggested that these variations were related closely to changes in the weather. This was the origin of the barometer. With the concept thus established that air had weight, Torricelli was able to predict that there must be less atmospheric pressure at higher altitudes. In 1646, Blaise Pascal, aided by a brother-in-law, carried a barometer to different levels of the 4,000-foot mountain Puy de Dôme, near Clermont, to take readings. The confirmation was that atmospheric pressure varied with altitude.

MERCURY THERMOMETER MEASUREMENTS

The mercury thermometer originated circa 1714–1742. Gabriel Daniel Fahrenheit, a German glassblower and instrument-maker working in Holland, was interested in improving the design of the thermometer, which had been in use for half a century. Known as the Florentine thermometer (because it was developed in the 1650s in Florence's Accademia del Cimento), this pioneering instrument depended on the expansion and contraction of alcohol within a glass tube. Alcohol expands rapidly with a rise in temperature, but not at an entirely regular speed of expansion. This made accurate readings difficult, as did the sheer technical problem of blowing glass tubes with very narrow and entirely consistent bores. By 1714, Fahrenheit had made great progress on the technical front, creating two separate alcohol thermometers that agreed precisely in their reading of temperature. In that year, he heard of the research of a French physicist, Guillaume Amontons, into the thermal properties of mercury. Mercury expands less than alcohol (about seven times less for the same rise in temperature), but it does so in a more regular manner. Fahrenheit saw the advantage of this regularity, and he used his glass-making skills to fashion a glass bore that accommodated the smaller rate of expansion. He constructed the first mercury thermometer, which subsequently became a standard. There remained the problem of how to calibrate the thermometer to show degrees of temperature. The only practical method was to choose two temperatures that could be established independently, mark them on the thermometer, and divide the intervening length of tube into a number of equal degrees. In 1701, Sir Isaac Newton had proposed the freezing point of water for the bottom of the scale and the temperature of the human body for the top end. Fahrenheit, accustomed to Holland's cold winters, wanted to include temperatures below the freezing point of water. He, therefore, accepted blood temperature for the top of his scale but adopted the freezing point of salt water for the lower extreme.

Measurement is conventionally done in multiples of 2, 3, and 4, so Fahrenheit split his scale into 12 sections, each of them divided into 8 equal parts. This gave him a total of 96 degrees, zero being the freezing point of brine and 96° as an inaccurate estimate of the average temperature of the human blood. Actual mean human body temperature is 98.6°. With his thermometer calibrated on these two points, Fahrenheit could take a reading for the freezing point (32°) and boiling point (212°) of water. In 1742, a Swede, Anders Celsius, proposed an early example of decimalization. His centigrade scale took the freezing and boiling temperatures of water as 0° and 100°. In English-speaking countries, this less complicated system took more than two centuries to be embraced. Even today, the Fahrenheit unit of temperature is more prevalent in some countries, such as the USA.

THE CHRONOMETER

The chronometer was developed circa 1714–1766. Two centuries of ocean travel made it important for ships on naval or merchant business to be able to calculate their positions accurately in any of the oceans in the world. With the help of the simple and ancient astrolabe, the stars would reveal latitude. But on a revolving planet, longitude is harder. It was essential to know what time it was before it could be determined what place it was. The importance of this was made evident in 1714 when the British government set up a Board of Longitude and offered a massive prize of £20,000 to any inventor who could produce a clock capable of keeping accurate time at sea. The terms were demanding. To win the prize, a chronometer had to be sufficiently accurate to calculate longitude within 30 nautical miles at the end of a journey to the West Indies. This meant that in rough seas, damp salty conditions, and sudden changes of temperature, the instrument must lose or gain not more than 3 seconds a day. This was a level of accuracy unmatched at the time by the best clocks. The challenge appealed to John Harrison, who was at the time a 21-year-old Lincolnshire carpenter with an interest in clocks. It was nearly 60 years before he won the money. Luckily, he lived long enough to collect it. By 1735, Harrison had built the first chronometer, which he believed approached the necessary standard. Over the next quarter-century, he replaced it with three improved models before formally undergoing the government's test. His innovations included bearings to reduce friction, weighted balances interconnected by coiled springs to minimize the effects of movement, and the use of two metals in the balance spring to cope with expansion and contraction caused by changes of temperature. Harrison's first "sea clock," in 1735, weighed 72 pounds and was 3 feet in all dimensions. His fourth, in 1759, was more like a watch, being circular and 5 inches in diameter. It was this version that underwent the sea trials. Harrison was at that time 67 years old. So, his son took the chronometer on its test journey to Jamaica in 1761. It was 5 seconds slow at the end of the voyage. The government argued that this might be a fluke and offered Harrison only £2,500. After further trials, and the successful building of a Harrison chronometer by another craftsman (at the huge cost of £450), the inventor was finally paid the full prize money in 1773.

Harrison proved in 1761 what was possible, but his chronometer was an elaborate and expensive way of achieving the purpose. It was in France, where a large prize was also on offer from the Académie des Sciences, that the practical chronometer of the future was developed. The French trial, open to all comers, took place in 1766 on a voyage from Le Havre in a specially commissioned yacht, the *Aurore*. The only chronometer ready for the test was designed by Pierre Le Roy. At the end of 46 days, his machine was accurate to within 8 seconds. Le Roy's timepiece was larger than Harrison's final model, but it was much easier to construct. It provided the pattern of the future. With further modifications from various sources over the next two decades, the marine chronometer emerged before the end of the 18th century. Using it in combination with the sextant, explorers traveling the world's oceans could then bring back accurate information of immense value to the makers of maps and charts of the world.

MEASUREMENT BY THE SEXTANT

The sextant originated between 1731 and 1757. The 18th-century search for a way to discover longitude was accompanied by refinements in the ancient method of establishing latitude. This had been possible since the 2nd century BC by means of the *astrolabe*. From the beginning of European voyages in the 15th century, practical improvements had been made to the astrolabe, mainly by providing more convenient calibrated arcs on which the user could read the number of degrees of the sun or a star above the horizon. The size of these arcs was defined in relation to the full circle. A quadrant (a quarter of the circle) showed 90°, a sextant 60°, and an octant 45°. The use of such arcs in conjunction with the traditional astrolabe is evident from a text dating back to 1555 that chronicled voyages to the West Indies. The author talked of "quadrant and astrolabe, instruments of astronomy." An important "innovative" improvement during the 18th century was the application of optical devices (mirrors and lenses) to promote the task of working out angles above the horizon. Slightly differing solutions arose from instrument makers in Europe and America, which competed during the early decades of the century. The one that prevailed, mainly because it was more convenient at sea, was designed as an octant in 1731 by John Hadley, an established English maker of reflecting telescopes. Hadley's instrument, like others designed by his contemporaries, used mirrors to bring any two points into alignment in the observer's sight-line. For the navigator at that time, these two points would usually be the sun and the horizon. To read the angle of the sun, the observer looked through the octant's eyepiece at the horizon and then turned an adjusting knob until the reflected sphere of the sun (through a darkened glass) was brought down to the same level. The double reflection meant that the actual angle of the sun above the horizon was twice that of the octant's arc of 45%. So, Hadley's instrument could read angles up to 90%. In 1734, Hadley added an improvement that became the standard by installing an additional level so that the horizontal could be found even if the horizon was not visible. In 1757, after Hadley's death, a naval captain proposed that the arc in the instrument be extended from 45° to 60°, making possible a reading up to 120°. With this, Hadley's octant became a sextant, and the instrument came into general use after that.

ANCIENT MEASUREMENT SYSTEMS IN AFRICA

Africa is home to the world's earliest known use of measuring and calculation, confirming the continent as the origin of both basic and advanced mathematics. Thousands of years ago, while parallel developments were going on in Europe, Africans were using rudimentary numerals, algebra, and geometry in daily life. This knowledge spread throughout the entire world after a series of migrations out of Africa, beginning around 30,000 BC, and later following a series of invasions of Africa by Europeans and Asians (1700 BC to the present). It is historically documented that early man migrated out of Africa to Europe and Asia. This feat of early travel and navigation could have been facilitated by the indigenous measurement systems of ancient Africa. The following sections recount measuring and counting in ancient Africa.

MEASUREMENT BY THE LEBOMBO BONE (PREHISTORIC)

The oldest known mathematical instrument is the Lebombo bone (see Figure 1.2), a baboon fibula used as a measuring device and so named for its location of discovery

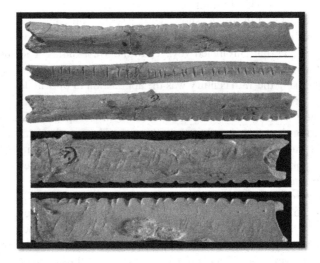

FIGURE 1.2   Image of Ancient Lebombo Bone for Ancient Measurement

in the Lebombo mountains of Swaziland. The device is at least 35,000 years old. Judging from its 29 distinct markings, it could have been either used to track lunar cycles or used as a measuring stick. It is rather interesting to note the significance of the 29 markings (roughly the same number as days in the lunar cycle, i.e., 29.531 days) on the baboon fibula, because it is the oldest indication that the baboon, a primate indigenous to Africa, was symbolically linked to Khonsu, who was also associated with time. The Kemetic god Djehuty ("Tehuti" or "Toth") was later depicted as a baboon or an ibis, which is a large tropical wading bird with a long neck and long legs. This animal symbolism is usually associated with the moon, math, writing, and science. The use of baboon bones as measuring devices had been continuous throughout all of Africa, suggesting that Africans always held the baboon as sacred and associated with the moon, math, and time.

MEASUREMENT BY THE ISHANGO BONE (PREHISTORIC)

The world's oldest evidence of advanced mathematics was also a baboon fibula that was discovered in the present-day Democratic Republic of Congo and dates to at least 20,000 BC. The bone is now housed in the Museum of Natural Sciences in Brussels. The Ishango bone is not merely a measuring device or tally stick, as some people erroneously suggest. The bone's inscriptions are clearly separated into clusters of markings that represent various quantities. When the markings are counted, they are all odd numbers, with the left column containing all prime numbers between 10 and 20, and the right column containing added and subtracted numbers. When both columns are calculated, they add up to 60 (nearly double the length of the lunar cycle). We recall that the number 60 also featured prominently in the development of early measuring devices in Europe.

BOARD GAME MEASUREMENTS OF GEBET'A AND MANCALA

Although the oldest known evidence of the ancient counting board game Gebet'a (or "Mancala" as it is more popularly known) comes from Yeha (700 BC) in Ethiopia, it was probably used in Central Africa many years prior to that. The game forced a player to strategically capture a greater number of stones than their opponent. The game usually consists of a wooden board with two rows of six holes each, and two larger holes at either end. However, in antiquity, the holes were more likely to be carved into stone, clay, or mud. More advanced versions found in Central and East Africa, such as the Omweso, Igisoro, and Bao, usually involve four rows of eight holes each. A variant of this counting game still exists today in the Yoruba culture of Nigeria. It is called the "Ayo" game, which tests the counting and tracking ability of players. A modern Yoruba Ayo game board (see Figure 1.3) has a row of six holes on each player's side, with a master counting hole above the row. This example and other similar artifacts demonstrate the handed-down legacy of ancient counting and measurements in Africa. These indigenous board games were the innovations of their respective eras.

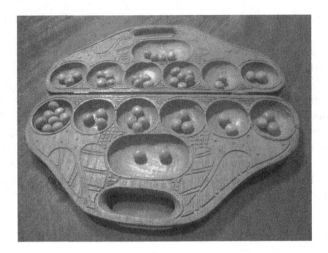

FIGURE 1.3   Nigerian Yoruba Ayo Game Board as a Numeric Counting Tool

MOSCOW PAPYRUS

Housed in Moscow's Pushkin State Museum of Fine Arts, the so-called "Moscow" papyrus was purchased by Vladimir Golenishchev some time in the 1890s. Written in hieratic from, perhaps, the 13th dynasty in Kemet, the name of ancient Egypt, the papyrus is one of the world's oldest examples of use of geometry and algebra. The document contains approximately 25 mathematical problems, including how to calculate the length of a ship's rudder, the surface area of a basket, the volume of a frustum (a truncated pyramid), and various ways of solving for unknowns. One feature of the papyrus is a number with a bar over it, which represents the unit fraction of 1/3. Fractions were important to the Egyptians because they allowed them to divide objects for construction (such as bricks for the pyramids) to coincide with the number of workers. There were tens of thousands of workers, so this task was mathematically imperative.

"RHIND" MATHEMATICAL PAPYRUS

Purchased by Alexander Rhind in 1858 AD, the so-called "Rhind" Mathematical Papyrus dates to approximately 1650 BC and is presently housed in the British Museum. Although some Egyptologists link this to the foreign Hyksos, this text was found during excavations at the Ramesseum in Waset (Thebes) in Southern Egypt, which never came under Hyksos rule. The first page contains 20 arithmetic problems, including addition and multiplication of fractions, and 20 algebraic problems, including linear equations. The second page shows how to calculate the volume of rectangular and cylindrical granaries, with pi (Π) estimated at 3.1605. There are also calculations for the area of triangles (slopes of a pyramid) and an octagon. The third page continues with 24 problems, including the multiplication of algebraic fractions, among others. Notably, these two papyri are about 200 years apart, and the Rhind

Papyrus is considerably more detailed. Using today's pace of discovery and invention, this same additional detail would have been achieved in 0.3 seconds. The accelerated pace of discovery is a key component of innovation, which will be explained in more detail in a subsequent chapter.

TIMBUKTU MATHEMATICAL MEASUREMENTS

Timbuktu in Mali is well-known as a hub of commerce in ancient times. Timbuktu is home to one of the world's oldest universities, Sankore, which had libraries full of manuscripts, mainly written in Ajami (African languages, such as Ajami and Hausa, were written in a script similar to Arabic) in the 1200s AD. When Europeans and Western Asians began visiting and colonizing Mali from the 1300s to the 1800s AD, Malians began to hide the manuscripts, as they were seen as heresy by certain religious groups visiting the region. Many of the scripts were mathematical and astronomical in nature. In recent years, as many as 700,000 scripts have been rediscovered and attest to the continuous knowledge of advanced mathematics, science, and measurements in Africa well before European colonization.

FUNDAMENTAL SCIENTIFIC EQUATIONS

In our modern time, we rely more and more on scientific equations to convey the values and properties of our tools and assets. This section presents some of the seminal and fundamental, theoretical scientific equations that have emerged over the centuries. Perhaps the most quoted and recognized in modern scientific literature is the equation of relativity developed by Albert Einstein.

EINSTEIN'S EQUATION

$$E = mc^2$$

The fundamental relationship connecting energy, mass, and the speed of light emerges from Einstein's theory of special relativity, published in 1905. Showing the equivalence of mass and energy, it may be the most famous and beautiful equation in all of modern science. Its power was graphically demonstrated less than four decades later with the discovery of nuclear fission, a process in which a small amount of mass is converted to a very large amount of energy, precisely in accord with this equation.

EINSTEIN'S FIELD EQUATION

$$R_{\mu\nu} - \frac{1}{2} g_{\mu\nu} R + \Lambda g_{\mu\nu} = 8\pi G T_{\mu\nu}$$

Einstein's elegant field equation, published in 1916, is the foundation for his theory of gravity, the theory of general relativity. The equation relates the geometrical curvature of space-time to the energy density of matter. The theory constructs an entirely new picture of space and time, out of which gravity emerges in the form of geometry and from which Newton's theory of gravity emerges as a limiting case. Einstein's field equation explains many features of modern cosmology, including the expansion of the universe and the bending of star light by matter, and it predicts black holes and gravitational waves. He introduced a cosmological constant in the equation, which he called his greatest blunder, but that quantity may be needed if, as recent observations suggest, the expansion of the universe is accelerating. A remaining challenge for physicists in the 21st century is to produce a fundamental theory uniting gravitation and quantum mechanics.

HEISENBERG'S UNCERTAINTY PRINCIPLE

$$\Delta x \Delta p \geq \frac{h}{2}$$

Werner Heisenberg's matrix formulation of quantum mechanics led him to discover in 1927 that an irreducible uncertainty exists when simultaneously measuring the position and momentum of a particle. Unlike classical mechanics, quantum mechanics requires that the more accurately the position of a particle is known, the less accurately its momentum is known, and vice versa. The magnitude of that irreducible uncertainty is proportional to Planck's constant.

SCHRÖDINGER EQUATION

$$i\hbar \frac{\partial \Psi}{\partial t} = H\Psi$$

In 1926, Erwin Schrödinger derived his non-relativistic wave equation for the quantum mechanical motion of particles such as electrons orbiting atoms. The probability density of finding a particle at a particular position in space is the square of the absolute value of the complex wave function, which is calculated from Schrödinger's equation. This equation accurately predicts the allowed energy levels for the electron in a hydrogen atom. With the use of modern computers, generalizations of this equation predict the properties of larger molecules and the behavior of electrons in complex materials.

DIRAC EQUATION

$$i\hbar \frac{\partial \Psi}{\partial t} = \left[c\vec{\alpha} \bullet (\vec{p} - \overline{A}) + \beta mc^2 + e\Phi \right] \Psi$$

In 1928, Paul Dirac derived a relativistic generalization of Schrödinger's wave equation for the quantum mechanical motion of a charged particle in an electromagnetic field. His marvelous equation predicts the magnetic moment of the electron and the existence of antimatter.

MAXWELL'S EQUATIONS

$$\vec{\nabla} \cdot \vec{D} = p$$

$$\vec{\nabla} \times \vec{H} = \vec{J} + \frac{\partial \vec{D}}{\partial t}$$

$$\vec{\nabla} \times \vec{E} + \frac{\partial \vec{B}}{\partial t} = 0$$

$$\vec{\nabla} \cdot \vec{B} = 0$$

The fundamental equations explaining classical electromagnetism were developed over many years by James Clerk Maxwell and finished in his famous treatise published in 1873. His classical field theory provides a refined framework for understanding electricity, magnetism, and the propagation of light. Maxwell's theory was a major achievement of 19th-century physics, and it contained one of the clues that were used years later by Einstein to develop special relativity. Classical field theory was also the springboard for the development of quantum field theory.

BOLTZMANN'S EQUATION FOR ENTROPY

$$S = k \ln W$$

Ludwig Boltzmann, one of the founders of statistical mechanics in the late 19th century, proposed that the probability for any physical state of a macroscopic system is proportional to the number of ways in which the internal state of that system can be rearranged without changing the system's external properties. When more arrangements are possible, the system is more disordered. Boltzmann showed that the logarithm of the multiplicity of states of a system, or its disorder, is proportional to its entropy, and the constant of proportionality is Boltzmann's constant k. The second law of thermodynamics states that the total entropy of a system and its surroundings always increases as time elapses. Entropy is a measure of the disorder of a system and is therefore designated a unit related to "a system's thermal energy per unit temperature that is unavailable for doing useful work" (Drake, 2022). According to NIST, the unit for entropy, designated S, is joules per kelvin (J/K). An entropy unit, e.u., is equal to one calorie per kelvin per mole, or 4.184 joules per kelvin per mole. Entropy can be considered the opposite of innovation. Innovation is making order out of chaos or organizing disparate tools and ideas into new systems and capabilities, while harnessing every last bit of energy for usefulness. Innovation can be an attempt to reduce entropy. Incidentally, in a satirical end to the scientist's methodical life, Boltzmann's equation for entropy is carved on his grave.

PLANCK–EINSTEIN EQUATION

$$E = h\nu$$

The simple relation between the energy of a light quantum and the frequency of the associated light wave first emerged in a formula discovered in 1900 by Max Planck. He was examining the intensity of electromagnetic radiation emitted by atoms in the walls of an enclosed cavity (a blackbody) at fixed temperature. He found that he could fit the experimental data by assuming that the energy associated with each mode of the electromagnetic field was an integral multiple of some minimum energy that is proportional to the frequency. The constant of proportionality, h, is known as Planck's constant. It is one of most important fundamental numbers in physics. In 1905, Albert Einstein recognized that Planck's equation implies that light is absorbed or emitted in discrete quanta, explaining the photoelectric effect and igniting the quantum mechanical revolution.

PLANCK'S BLACKBODY RADIATION FORMULA

$$u = \frac{8\pi h}{c^3} \nu^3 \left[e^{\frac{h\nu}{kT}} - 1 \right]^{-1}$$

In studying the energy density of radiation in a cavity, Max Planck compared two approximate formulas, one for low frequency and one for high frequency. In 1900, using an ingenious extrapolation, he found his equation for the energy density of black body radiation, which reproduced experimental results. Seeking to understand the significance of his formula, he discovered the relation between energy and frequency known as the Planck–Einstein equation.

HAWKING EQUATION FOR BLACK HOLE TEMPERATURE

$$T_{BH} = \frac{hc^3}{8\pi GMk}$$

Using insights from thermodynamics, relativist quantum mechanics, and Einstein's gravitational theory, Stephen Hawking predicted in 1974 the surprising result that gravitational black holes, which are predicted by general relativity, would radiate energy. His formula for the temperature of the radiating black hole depends on the gravitational constant (Planck's constant), the speed of light, and Boltzmann's constant. While Hawking's radiation theory remains to be observed, his formula provides a tempting glimpse of the insights that will be uncovered in a unified theory combining quantum mechanics and gravity. Perhaps innovation in space will pave the way to designing an experimental method to observe this incredible phenomenon.

NAVIER–STOKES EQUATION FOR A FLUID

$$\rho\frac{\partial \vec{v}}{\partial t}+\rho\left(\vec{\upsilon}\cdot\vec{\nabla}\right)\vec{\upsilon}=-\vec{\nabla}p+\mu\nabla^2\vec{\upsilon}+\left(\lambda+\mu\right)\vec{\nabla}\left(\vec{\nabla}\cdot\vec{\upsilon}\right)+\rho\vec{g}$$

The Navier–Stokes equation was derived in the 19th century from Newtonian mechanics to model viscous fluid flow. Its nonlinear properties make it extremely difficult to solve, even with modern analytic and computational techniques. However, its solutions describe a rich variety of phenomena, including turbulence.

LAGRANGIAN FOR QUANTUM CHROMODYNAMICS

$$L_{QDC}=-\frac{1}{4}F_a^{\mu\nu}\cdot F_{a\mu\nu}+\sum_f\overline{\Psi}\left[i\slashed{\partial}-g\slashed{A}_a t_a-m_f\right]\Psi_f$$

Relativistic quantum field theory had its first great success with quantum electrodynamics, which explains the interaction of charged particles with the quantized electromagnetic field. Exploration of non-Abelian gauge theories led next to the spectacular unification of electromagnetic and weak interactions. Then, with insights developed from the quark model, quantum chromodynamics was developed to explain the strong interactions. This theory predicts that quarks are bound more tightly together as their separation increases, which explains why individual quarks are not seen directly in experiments. The standard model, which incorporates strong, weak, and electromagnetic interactions in a single quantum field theory, describes the interaction of quarks, gluons, and leptons and has achieved remarkable success in predicting experimental results in elementary particle physics.

Bardeen–Cooper–Schrieffer (BCS) Equation for Superconductivity

$$T_c=1.13\Theta e^{-\frac{1}{N(0)V}}$$

Superconductors are materials that exhibit no electrical resistance at low temperatures. In 1957, John Bardeen, Leon N. Cooper, and J. Robert Schrieffer applied quantum field theory with an approximate effective potential to explain this unique behavior of electrons in a superconductor. The electrons were paired and moved collectively without resistance in the crystal lattice of a superconducting material. The BCS theory and its later generalizations predict a wide variety of phenomena that agree with experimental observations and have many practical applications. John Bardeen's contributions to solid state physics also include inventing the transistor, made from semiconductors, with Walter Brattain and William Shockley in 1947. Talk about innovation: Our entire planet is benefitting from this particular invention in myriad ways.

Josephson Effect

$$\frac{d(\Delta\varphi)}{dt} = \frac{2eV}{h}$$

In 1962, Brian Josephson made the remarkable prediction that electric current could flow between two thin pieces of superconducting material separated by a thin piece of insulating material (called a Josephson junction) without application of a voltage. Using the BCS theory of superconductivity, he also predicted that if a voltage difference were maintained across the junction, there would be an alternating current with a frequency related to the voltage and Planck's constant. The presence of magnetic fields influences the Josephson effect, allowing it to be used to measure very weak magnetic fields approaching the microscopic limit set by quantum mechanics. Incidentally, superconductivity is another marvelous innovation, as its power can be harnessed to construct lightweight and small systems at atmospheric pressures that balance the required lower temperatures according to the Ideal Gas Law: PV = nRT.

Fermat's Last Theorem

$$x^n + y^n = z^n$$

While studying the properties of whole numbers, or integers, the French mathematician Pierre de Fermat wrote in 1637 that it is impossible for the cube of an integer to be written as the sum of the cubes of two other integers. More generally, he stated that it is impossible to find such a relation between three integers for any integral power greater than two. He went on to write a tantalizing statement in the margin of his copy of a Latin translation of Diophantus' *Arithmetica*: "I have a truly marvelous demonstration of this proposition, which this margin is too narrow to contain." It took over 350 years to prove Fermat's simple conjecture. The feat was achieved by Andrew Wiles in 1994 with a "tour de force" proof of many pages using newly developed techniques in number theory. It is noteworthy that many researchers, mathematicians, and scholars toiled for almost four centuries before a credible proof of Fermat's last theorem was found. Published references on Fermat's last theorem are plentiful. Interested readers may look them up.

FUNDAMENTAL METHODS OF MEASUREMENT

There are two basic methods of measurement:

1. *Direct comparison* with either a primary or a secondary standard
2. *Indirect comparison* with a standard through the use of a calibrated system

Direct comparison. How do you measure the length of a cold-rolled bar? You probably use a steel tape. You compare the bar's length with a standard. The bar is so many feet long because that many units on your standard have the same length as the bar. You have determined this by making a direct comparison. Although you do not have access to the primary standard defining the unit, you manage very well with a secondary standard. Primary measurement standards have the lowest amount of uncertainty compared with the certified value and are traceable directly to the SI. Secondary standards, on the other hand, are derived by ratifying their accuracy through direct comparison with a primary standard.

In some respects, measurement by direct comparison is quite common. Many length measurements are made in this way, using a ruler or tape measure, a tool considered to be a primary standard. In addition, time of day is usually determined by glancing at one's watch (indirect comparison) and using the watch as a secondary standard. The watch goes through its double cycle (two 12-hour periods) in synchronization with the earth's rotation. Although, in this case, the primary standard is available to everyone (looking up in the sky at the sun from the earth), the watch is more convenient because it works on cloudy days, indoors, outdoors, in daylight, and in the dark (at night). It is also more precise. That is, its resolution is better. In addition, if well-regulated, the watch is more accurate, because the earth does not rotate at a uniform speed. It is seen, therefore, that in some cases, a secondary standard is actually more useful than the primary standard.

Measuring by direct comparison implies stripping the measurement problem to its barest essentials. However, the method is not always the most accurate or the best. The human senses are not equipped to make direct comparisons of all quantities with equal facility. In many cases, they are not sensitive enough. We can make direct length comparisons using a steel rule with a level of precision of about 0.01 inch. Often, we wish for greater accuracy, in which case we must call for additional assistance from some calibrated measuring system.

Indirect comparison. While we can do a reasonable job through direct comparison of length, how well can we compare masses, for example? Our senses enable us to make rough comparisons. We can lift a pound of meat and compare its effect with that of some unknown mass. If the unknown is about the same weight, we may be able to say that it is slightly heavier, or perhaps, not quite as heavy as our "standard" pound, but we could never be certain that the two masses were the same, even say within 1 ounce. Our ability to make this comparison is not as good as it is for the displacement of the mass. Our effectiveness in coming close to the standard is related to our ability to "gauge" the relative impacts of mass on our ability to displace the mass. This brings to mind the common riddles of "Which weighs more? A pound of feathers or a pound of stones?" Of course, both weigh the same with respect to the standard weight of "pound."

In making most engineering measurements, we require the assistance of some form of measuring system, and measurement by direct comparison is less general than measurement by indirect comparison.

GENERALIZED MECHANICAL MEASURING SYSTEM

Most mechanical measurement systems (Beckwith and Buck, 1965) fall within the framework of a generalized arrangement consisting of three stages, as follows:

Stage I: A detector-transducer stage
Stage II: An intermediate modifying stage
Stage III: The terminating stage, consisting of one or a combination of an indi-
    cator, a recorder, or some form of controller

Each stage is made up of a distinct component or grouping of components that perform required and definite steps in the measurement. These may be termed *basic elements*, whose scope is determined by their functioning rather than their construction. First stage detector-transducer: The prime function of the first stage is to detect or to sense the input signal. This primary device must be sensitive to the input quantity. At the same time, ideally, it should be insensitive to every other possible input. For instance, if it is a pressure pickup, it should not be sensitive to, say, acceleration; if it is a strain gauge, it should be insensitive to temperature; or if a linear accelerometer, it should be insensitive to angular acceleration, and so on. Unfortunately, it is very rare indeed to find a detecting device that is completely selective. As an example of a simple detector-transducer device, consider an automobile tire pressure gauge. It consists of a cylinder and a piston, a spring resisting the piston movement, and a stem with scale divisions. As the air pressure bears against the piston, the resulting force compresses the spring until the spring and air forces are balanced. The calibrated stem, which remains in place after the spring returns the piston, indicates the applied pressure. In this case, the piston–cylinder combination along with the spring makes up the detector-transducer. The piston and cylinder form one basic element, while the spring is another basic element. The piston–cylinder combination, serving as a force-summing device, senses the pressure effect, and the spring transduces it into the displacement. Realistically, not all measurements we encounter in theory and practice are of transduce-able mechanical settings. Measurements, thus, can take more generic paths of actualization. Figure 1.4 shows a generic measurement loop revolving around variable identification, actual measurement, analyzing the measurement result, interpreting the measuring in the context of the prevailing practical application, and implementing the measurement for actionable decisions. In each stage of the loop, communication is a central requirement. Communication can be in the form of a pictorial display, a verbal announcement, or a written dissemination. A measurement is not usable unless it is communicated in an appropriate form and at the appropriate time.

MEASUREMENT SCALES AND DATA TYPES

Every decision requires data collection, measurement, and analysis. In practice, we encounter different types of measurement scales depending on the particular items of interest. Data may need to be collected on decision factors, costs, performance levels, outputs, and so on. The different types of data measurement scales that are applicable are presented as follows.

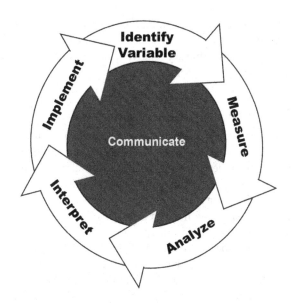

FIGURE 1.4 Generic Measurement Loop

NOMINAL SCALE OF MEASUREMENT

Nominal scale is the lowest level of measurement scales. It classifies items into categories. The categories are mutually exclusive and collectively exhaustive. That is, the categories do not overlap, and they cover all possible categories of the characteristics being observed. For example, in the analysis of the critical path in a project network, each job is classified as either critical or not critical. Gender, type of industry, job classification, and color are examples of measurements on a nominal scale.

ORDINAL SCALE OF MEASUREMENT

Ordinal scale is distinguished from a nominal scale by the property of order among the categories. An example is the process of prioritizing project tasks for resource allocation. We know that first is above second, but we do not know how far above. Similarly, we know that better is preferred to good, but we do not know by how much. In quality control, the ABC classification of items based on the Pareto distribution is an example of a measurement on an ordinal scale.

INTERVAL SCALE OF MEASUREMENT

Interval scale is distinguished from an ordinal scale by having equal intervals between the units of measurement. The assignment of priority ratings to project objectives on a scale of 0 to 10 is an example of a measurement on an interval scale. Even though an objective may have a priority rating of zero, it does not mean that the objective has absolutely no significance to the project team. Similarly, the scoring of zero on an examination does not imply that a student knows absolutely nothing about the materials covered by the examination. Temperature is a good example of an item that is measured on an interval scale. Even though there is a zero point on the temperature scale, it is an arbitrary relative measure. Other examples of interval scale are IQ measurements and aptitude ratings.

RATIO SCALE MEASUREMENT

Ratio scale has the same properties as an interval scale, but with a true zero point. For example, an estimate of a zero-time unit for the duration of a task is a ratio scale measurement. Other examples of items measured on a ratio scale are cost, time, volume, length, height, weight, and inventory level. Many of the items measured in engineering systems will be on a ratio scale.

Another important aspect of measurement involves the classification scheme used. Most systems will have both quantitative and qualitative data. Quantitative data require that we describe the characteristics of the items being studied numerically. Qualitative data, on the other hand, are associated with attributes that are not measured numerically. Most items measured on the nominal and ordinal scales will normally be classified into the qualitative data category, while those measured on the interval and ratio scales will normally be classified into the quantitative data category. The implication for engineering system control is that qualitative data can lead to bias in the control mechanism because qualitative data are subject to the personal views and interpretations of the person using the data. As much as possible, data for an engineering systems control should be based on a quantitative measurement. Figure 1.5 illustrates the four different types of data classification. Notice that temperature is included in the "relative" category rather than the "true zero" category. Even though there are zero temperature points on the common temperature scales

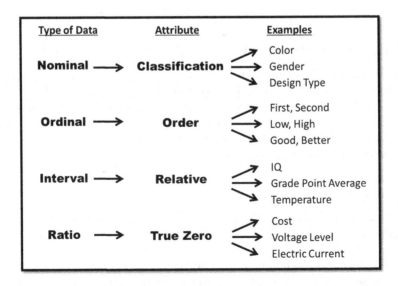

FIGURE 1.5 Four Primary Types of Data Measurement Scales

(i.e., Fahrenheit, Celsius, and Kelvin), those points are experimentally or theoretically established. They are not true points as one might find in a counting system.

COMMON UNITS OF MEASUREMENT

Some common units of measurement include the following:

Acre: An area of 43,560 square feet.

Agate: 1/14 inch (used in printing for measuring column length).

Ampere: Unit of electric current.

Astronomical (A.U.): 93,000,000 miles; the average distance of the earth from the sun (used in astronomy).

Bale: A large bundle of goods. In the USA, approximate weight of a bale of cotton is 500 pounds. Weight of a bale may vary from country to country.

Board Foot: 144 cubic inches (12 by 12 by 1 used for lumber).

Bolt: 40 yards (used for measuring cloth).

BTU: British thermal unit; amount of heat needed to increase the temperature of 1 pound of water by 1 degree Fahrenheit (252 calories).

Carat: 200 milligrams or 3,086 troy; used for weighing precious stones (originally the weight of a seed of the carob tree in the Mediterranean region). *See also Karat.*

Chain: 66 feet; used in surveying (1 mile = 80 chains).

Cubit: 18 inches (derived from distance between elbow and tip of middle finger).

Decibel: Unit of relative loudness.

Freight Ton: 40 cubic feet of merchandise (used for cargo freight).

Gross: 12 dozen (144).

Hertz: Unit of measurement of electromagnetic wave frequencies (measures cycles per second).

Hogshead: Two liquid barrels or 14,653 cubic inches.

Horsepower: The power needed to lift 33,000 pounds a distance of 1 foot in 1 minute (about 1½ times the power an average horse can exert); used for measuring power of mechanical engines.

Karat: A measure of the purity of gold. It indicates how many parts out of 24 are pure. 18 karat gold is ¾ pure gold.

Knot: Rate of speed of 1 nautical mile per hour; used for measuring speed of ships (not distance).

League: Approximately 3 miles.

Light year: 5,880,000,000,000 miles; distance traveled by light in 1 year at the rate of 186,281.7 miles per second; used for measurement of interstellar space.

Magnum: Two-quart bottle; used for measuring wine.

Ohm: Unit of electrical resistance.

Parsec: Approximately 3.26 light-years or 19.2 trillion miles; used for measuring interstellar distances.

Pi (π): 3.14159265+; the ratio of the circumference of a circle to its diameter.

Pica: 1/6 inch or 12 points; used in printing for measuring column width.

Pipe: Two hogsheads; used for measuring wine and other liquids.

Point: 0.013837 (approximately 1/72 inch or 1/12 pica); used in printing for measuring type size.

Quintal: 100,000 grams or 220.46 pounds avoirdupois.

Quire: 24 or 25 sheets; used for measuring paper (20 quires is one ream).

Ream: 480 or 500 sheets; used for measuring paper.

Roentgen: Dosage unit of radiation exposure produced by X-rays.

Score: 20 units.

Span: 9 inches or 22.86 cm; derived from the distance between the end of the thumb and the end of the little finger when both are outstretched.

Square: 100 square feet; used in building.

Stone: 14 pounds avoirdupois in Great Britain.

Therm: 100,000 BTUs.

Township: US land measurement of almost 36 square miles; used in surveying.

Tun: 252 gallons (sometimes larger); used for measuring wine and other liquids.

Watt: Unit of power.

Apart from the total and average of a data set, the other data properties of interest for innovation metrics and assessment include mode, median, range, variance, deviation, quartiles, and percentiles of the data set. The median is a position measure because its value is based on its position in a set of observations. Other measures of position are *quartiles* and *percentiles*. There are three quartiles, which divide a set of data into four equal categories. The first quartile, denoted Q_1, is the value below which one-fourth of all the observations in the data set fall. The second quartile, denoted, Q_2, is the value below which two-fourths or one-half of all the observations in the data set fall. The third quartile, denoted Q_3, is the value below which three-fourths of the observations fall. The second quartile is identical to the median. It is technically incorrect to talk of the fourth quartile, because that will imply that there is a point within the data set below which all the data points fall: a contradiction! A data point cannot lie within the range of the observations and at the same time exceed all the observations, including itself.

The concept of percentiles is similar to the concept of quartiles, except that reference is made to percentage points. There are 99 percentiles that divide a set of observations into 100 equal parts. The X percentile is the value below which X percent of the data fall. The 99 percentile refers to the point below which 99 percent of the observations fall. The three quartiles discussed previously are regarded as the 25th, 50th, and 75th percentiles. It would be technically incorrect to talk of the 100th percentile. In performance ratings, such as on an examination or product quality level, the higher the percentile of an individual or product, the better. In many cases, recorded data are classified into categories that are not indexed to numerical measures. In such cases, other measures of central tendency or position will be needed. The mode is defined as the value that has the highest frequency in a set of observations. When the recorded observations can be classified only into categories, the mode can be particularly helpful in describing the data. Given a set of K observations (e.g., revenues), $X_1, X_2, X_3, \ldots, X_K$, the mode is identified as that value that occurs more than any other value in the set. Sometimes, the mode is not unique in a set of observations.

The results of data analysis can be reviewed directly to determine where and when project control actions may be needed. The results can also be used to generate control charts.

CONCLUSIONS

Human interfaces with measurement systems and tools date back to ancient times. Even in modern times, we must recognize past approaches to measurements, we must appreciate present practices of measurements, and we must be prepared for future developments of measurements for the purpose of driving and advancing innovation. To bring this point home, the Angbuilgu is a sundial that was used during the Josean Dynasty in Korea. The name "Angbuilgu" means "upward-looking kettle that catches the shadow of the sun." The 13 horizontal lines mark the 24 periods of seasonal change from the winter solstice to the summer solstice and allow the season of the year to be determined. The vertical lines indicate time, or where one is in time. The device is aligned to face the North Star, and it is marked with pictures of animals (rather than letters) for the sake of the common people, who could not read. The sundial tool was first made in 1434, but the later Angbuilgu was created in the latter half of the 17th century. This distinctive, historical tool that advanced the telling of time beyond anything that had come before it demonstrates that innovation by humans has been changing the world for generations. Now is the time to think of developing, communicating, understanding, and utilizing measurements for innovation.

REFERENCES

Badiru, A. B. and LeeAnn Racz (2016). *Handbook of Measurements: Benchmarks for Systems Accuracy and Precision*, Taylor & Francis CRC Press, Boca Raton, FL.

Badiru, A. B. and Tina Kovach (2012). *Statistical Techniques for Project Control*, Taylor & Francis CRC Press, Boca Raton, FL.

Badiru, A. B., Oye Ibidapo-Obe, and B. J. Ayeni (2012). *Industrial Control Systems: Mathematical and Statistical Models and Techniques*, Taylor & Francis CRC Press, Boca Raton, FL.

Beckwith, T. G. and N. L. Buck (1965). *Mechanical Measurements*, Addison-Wesley Publishing Company, Reading, MA.

Drake, G. W. F. (2022). *Entropy*, Encyclopedia Britannica. https://www.britannica.com/science/entropy-physics (accessed April 23, 2023).

HistoryWorld. (2014). http://www.historyworld.net/wrldhis/PlainTextHistories.asp?historyid=ac07 (accessed December 24, 2014).

Morris, A. S. (1997). *Measurement and Calibration Requirements for Quality Assurance to ISO 9000*, John Wiley & Sons, Inc., New York.

NIST. (1974). *NBS Special Publication; 304A (Stock Number 003-003-03501-7), September 1974*, U.S. Department of Commerce Technical Administration National Institute of Standards and Technology, Washington, DC.

Shillito, M. L. and D. J. De Marle (1992). *Value: Its Measurement, Design, and Management*, John Wiley & Sons, Inc., New York.

TANF. (2014). *Restoring Africa's Lost Legacy*, TA Neter Foundation. http://www.taneter.org/math.html (accessed December 23, 2014).

2 Defining Matrix and Metrics for Innovation

THE NEED FOR INNOVATION METRICS

To put things in perspective for why metrics are needed, we consider the following quote:

> I think innovation as a discipline needs to go back and get rethought and revived. There are so many models to talk about innovation, there are so many typologies of innovation, and you have to find a good innovation metric that truly captures the innovation performance of a company.
>
> **Indra Nooyi**

As pervasive and ubiquitous as innovation has become, we should have metrics for it, albeit with varying levels of affirmation and believability, particularly in diverse organizational settings. Without metrics, one innovation is just as good as any other innovation. The early innovations in warfare, such as a gun versus a sword, paved the way for many later technological innovations, thus providing a metric of innovation success. Other influential innovations over the ages include the umbrella, the typewriter, and the printing press. Nowadays, enmeshing technology, innovation, and partnerships can improve the satisfaction of established metrics, rubrics, and standards.

The term "innovation" has become the word of the day in business, industry, government, and the military. Everyone talks about innovation, but very few appreciate the mission-dependent implications. Nor does anyone present the metrics by which innovation can be assessed or affirmed. In one national advertisement, a university claims, "No one does innovation like we do." By what metrics of "yardsticking" did they arrive at this conclusion? That is a bold statement that is not backed by any measurable yardstick. This is another basis for presenting this book's focus on developing metrics by which innovation can be proclaimed.

Even the proponents of innovation often wonder what innovation implies. For this reason, we consider alternate definitions, views, and concepts underpinning innovation. Innovation is not necessarily a tangible product that can be physically appraised. Unfortunately, this is the default assumption of those who tout innovation. Such a default view of innovation, as an end product, misses the opportunity to recognize the necessary interplay of technology, people, and process.

DOI: 10.1201/9781003403548-2

In an April 2023 pronouncement about the investments the State of Ohio (USA) is making in workforce development, US Senator Sherrod Brown is quoted as saying: "We unleash more American innovation when we nurture Ohio talent" (McClory, 2023). This proclamation directly links innovation to national advancement. Many other governmental and political entities around the world are making similar statements that attach economic development aspirations to investments in innovation. How and when do we know that the proclaimed innovation is having the expected impacts? Standards and metrics are important for this purpose. The following list presents fodder for readers' consideration:

- Qualitative metrics
- Quantitative metrics
- Analogous metrics
- Subjective metrics
- Standards for innovation metrics
- Industry standard
- Government standards
- Consensus standards
- Social standards
- Market standards
- International standards
- Safety standards
- Legal standards
- Regulatory standards
- Economic standards
- Environmental standards

ITERATIVE AND DISRUPTIVE INNOVATION

David Bottiau, in his blog (Bottiau, 2018), defines iterative and disruptive innovation as narrated in the following:

- Iterative innovation: Start with something you know and go beyond.

Iterative innovation is the ability to improve what's currently available. We have a lot of examples to demonstrate what iterative innovation is. For example, in the food industry (restaurant), fast-foods are an alternative to existing restaurants, but they are based on the same principle: Deliver foods in a place. There is also a varying number of differentiation factors. The most common factors are time (delay), cost, and quality. Indeed, a fast-food is generally cheaper and faster to deliver food than a traditional restaurant. Iterative innovation is more common than disruptive innovation.

- Disruptive innovation: Start from scratch and break the rules.

Disruptive innovation appears when you completely change habits. Again, if we use the food industry example, we can try to find a way that people have not experimented with yet, a way that people could be interested in. An example of disruptive innovation in the food industry could be food injection. When we are working on innovation, customers are more willing to accept the change if it is an iterative innovation, because they are already familiar with something similar.

BREAKING RULES FOR INNOVATION: THE CASE OF OCEANGATE

On the subject of breaking rules for the sake of innovation, the tragic end of the OceanGate Titanic exploration submersible in June 2023 comes to mind. The inventor and explorer, Stockton Rush, was quoted as bragging that he broke a few rules in the pursuit of his inventive innovation of building the Titanic exploration submersible, named "The Titan Submersible." Rush claimed to have violated technical and operational rules in the design of the sub, but with logic and good engineering behind his approach. This is one example where the structured approach of the DEJI Systems Model, covered later in this book, could have helped OceanGate rationalize through the stages of the sub **Design**, **Evaluation**, **Justification**, and **Integration**. The submersible was basically experimental and had not been *evaluated* and approved by any ocean-faring regulatory agency. Thus, there was a catastrophic violation, not only of several safety policies put in place by qualified professionals to navigate the seas, but also of the second phase of the DEJI Systems Model. No universal standards and metrics were adhered to in building and operating the submersible. Without *EVALUATE*, the submersible underwent a terminal implosion, killing all five people on board. This OceanGate example will make a good systems engineering case study in the coming years. Many of the "ilities" of systems engineering are applicable in this case. Of particular concern is the lack of accountability in the face of dire odds.

Standards and metrics, if followed, can facilitate the pursuit of innovation with a deep purpose, albeit with an abundance of caution. Stockton Rush claimed that it would take too long for innovation to be assessed. Thus, his breaking of innovation rules was justified. This attitude flies in the face of accountability. The resulting consequence was a tragic end for the Titan submersible.

ITERATIVE INNOVATION

Iterative innovation has permeated human existence for generations. Many landmark innovations of the past were creatively iterated, but their function has been so deeply integrated into our daily lives that those improvements went unnoticed. Examples include the umbrella, the gun, the printing press, and the typewriter. A case example is the inverted or reverse umbrella that opens up outwardly so that those exiting a vehicle are protected upon opening a vehicle's door. This is

TABLE 2.1

Categories of Invention and Innovation

Invention	Innovation	Product
New	New	First
Existing	Improvement, but didn't change the world	Same
Existing	Yes, changed the world	Upgraded

obviously an iterative innovation of the conventional design of umbrellas, and a brilliant one at that, but it certainly didn't make mainstream news. What is interesting is that if one or several iterative innovations disappeared from our lives, there would be the proverbial "mayhem in the streets." On this note, a discussion of this topic would not be complete without mentioning the most compelling iterative innovation of the modern day: the telephone, or should we now say, the cellphone, or due to even more iterative innovation, the smartphone. Iterative and disruptive innovation therefore lead us to develop our first metric: Has innovation occurred or not? One could call it new innovation versus existing innovation. The metric is built on the premise that the product representing the innovation either existed before or is an entirely new invention. Sometimes, inventions of the past are re-engineered, re-enter the consumer market, and are branded as innovation. But by what metric do we assign the term "innovation" to products, especially if they are merely improvements to an existing product? Therein, again, lies the premise of this book (see Table 2.1). An appropriate question is whether or not a new innovation leads to a new invention.

EFFICIENCY AND INNOVATION

Marchet et al. (2017) present a compelling reason to assess innovation in the context of efficiency. The authors address an analytical process of assessing efficiency and innovation in the 3PL (Third-party Logistics) industry, based on an Italian case study. The 3PL industry is facing both important growth rates and increasing competitive pressure. Providers are required to continuously sustain a more and more competitive cost structure (i.e., efficiency) and develop capabilities to improve their services (i.e., innovation). Hence, the evaluation of these key success factors is considered a key issue. Marchet et al. (2017) present a quantitative analysis of 71 Italian 3PL providers by using Data Envelopment Analysis to jointly assess efficiency and innovation. Through a case study, the authors corroborate the quantitative results by investigating the strategies of best-in-class companies. Their results allowed them to identify 13 3PL providers as efficiency leaders and 6 as leaders from both the efficiency and the innovation metrics. The input composition indicates a diversification of the business models. The paper's conclusion presents a breakdown of the analysis by size and industry focus, along with empirical evidence on the strategies enhancing efficiency and innovation. The theme of the paper fits the premise of this book with respect to the need for a rigorous assessment and affirmation of innovation.

INNOVATION CAPABILITY

The term "innovation" refers to the capacity of firms to find solutions to existing problems and respond to challenges in the market (Badiru, 2023). Previous research has widely discussed innovation as one of the significant strategic benefits obtained from using information systems (Badiru, 2023). In line with Badiru (2023), innovation capability is defined as "a firm's ability to generate, accept, and implement new ideas, processes, products, or services." Information technology (IT) capabilities focus on deploying IT-related resources within the organization to improve firms' innovative capability (Badiru, 2023). For instance, Badiru (2023) argued that use of IT capabilities increases the firms' performance through assisting in innovative activities.

It is clear that several interconnected issues influence whether enhanced business analytics (BA) capabilities influence a firm's performance (Badiru, 2023). We argue that while BA can impact the quality of information in an organization, **innovation capability** (*the ability of an organization to perform innovative practices*) is equally important (Badiru, 2023). Both then improve the firm's agility, specified as the ability to sense and react to opportunities and threats with ease, speed, and dexterity (Badiru, 2023).

While firms can possess innovation capability, it is the people animating that capability that truly realize innovation for their employers. In this vein, it makes great sense to develop common measurements for innovation plans and communicate those plans as goals and metrics, such as Key Performance Indicators (KPIs), so that everyone is dancing to the same music sheet. Incidentally, a proven method for efficient communication is offered with the Triple C model of systems management. As with every other corporate endeavor, when people are properly prepared and expectations have been set, successful outcomes have a much greater chance of being realized. Additionally, while firms may possess innovation capabilities, so must their people. This book will suggest several human capabilities that are necessary to achieve firm innovation.

Innovation is a process rather than a product. Any methodology (of people and technology) designed to enhance that process is essentially facilitating innovation. Many people who flaunt innovation cannot even define what it is. Likewise, those who are expected to embrace innovation have no idea what it is. So, a basic understanding of innovation is essential. The definitions presented in Table 2.2 convey the various alternate views of innovation (Badiru, 2020).

Embedded across all these alternate definitions are different elements of importance depending on the audience. There are innovation proponents, advocates, sponsors, supporters, sponsors, developers, leaders, facilitators, and observers, a veritable

TABLE 2.2

Templates and Definitions of Innovation

Definition	Key Elements	Implication
"Innovation is a new way of doing something, a literal change to the way we live and operate, never to return to the old ways, and necessarily a valuable improvement and nothing less."	New pathway	Future focus
"Innovation is the methodology of managing, allocating, and timing organizational technology tools, workforce assets, work processes to achieve a given output in an efficient and expedient manner."	Methodology, managing, resources, efficiency	Output, expediency
"Innovation is creatively re-engineering solutions, for problems which may not yet exist, by actualizing new ideas into valuable processes, services, or goods."	Creativity, re-engineering, problem speculation	Solution, value-adding, services, goods
"Innovation is achieving a goal using novel means."	Goal, novel	Achievement
"The process of innovation is the management of resources to drive a new or unexpected result."	Process, result, management, resources	Unexpected, driving
"Innovation is a collection of principles to help facilitate problem-solving, looking at both traditional as well as non-traditional approaches."	Collection, principles, facilitate	Non-traditional
"Innovation is the project management process of employing novel resources and/ or methods to produce a product or a service that is more useful to a current or novel use case."	Project management, novel, resources, product, service	Use case, useful, current and future

pageant of stakeholders regarding innovation. Just as each would define innovation differently, each would present alternate factors for what impedes innovation, what facilitates innovation. and how best to drive innovation. These alternating viewpoints are summarized in the following.

THREATS TO INNOVATION

Just like any organizational asset, innovation is subject to threats and opportunities. The following are some selected impediments to innovation. The more the impediments are recognized, the more likely they can be mitigated within the framework of applicable metrics, rubrics, and standards. Similarly, the more the facilitators of innovation are identified, the more readily they can be leveraged.

1. What are impediments to innovation?
   a. Fear of taking risks
   b. Organizational momentum
   c. Stove-piping
   d. Customer isolation
   e. Lack of leadership
   f. Lack of vision – seeing the art of the possible
   g. Lack of sufficient time and poor time management
   h. Lack of stakeholder involvement
   i. Lack of structure communication
   j. Lack of innovative practices and understanding built into organizational culture
   k. Excessive risk aversion – fear of failure
   l. Lack of creativity
   m. Lack of diversity of thought
   n. Lack of funds
   o. Burdensome regulations
   p. Immovable bureaucracy (e.g., restrictive sources of funding)
   q. Status quo culture
   r. Ignorance about users' satisfaction with current products or services
   s. Fear of technological deviation
   t. Too busy to innovate
   u. Lack of "top-cover" from leadership
   v. Poor collaborating environment
   w. Insufficient resources (tools, budget, personnel)
   x. Organizational quagmire
   y. Customer isolation
   z. Fear of stepping out of the norm
   aa. Prior culture of poor cooperation and/or communication
2. What are facilitators of innovation?
   a. Communicate, communicate, and communicate again
   b. Top-down support and removal of "red tape"
   c. Integrate teams across organization, education, and training
   d. Talk with the customer
   e. Be willing to walk away
   f. Learn from failure
   g. Set SMART goals (Specific, Measurable, Aligned, Realistic/Achievable, and Timed)
   h. Hold people accountable for established goals
   i. Make innovation a priority
   j. Define risk boundaries

    k. Solicit feedback/listen
    l. Apply lessons learned
    m. Endorse training and promote teamwork
    n. Reward accomplishments; incentivize
    o. Top-down involvement (walk the walk, not just talk the talk)
    p. Get buy-in at the lowest levels
    q. Hire diversity to promote diversity of thought
    r. Create space and time for the pursuit of innovation, time to think
    s. Motivate employees with a clear vision
    t. Leadership employing change management
    u. Psychological safety to allow failure
    v. Failure learning via reporting/analysis/synthesis
    w. Create close interaction with various potential users and customers
    x. Institute problem-solving process
    y. Study the elements of innovation within the organization
    z. Embrace novel ideas, pursue, and reward

Facilitating innovation can best be summed up by Andre Gide:

> One does not discover new lands without consenting to lose sight of the shore for a very long time.

SMART INNOVATION

The principles of the SMART goals mentioned earlier can be applied directly to innovation under the premise of SMART Innovation. Innovation performance elements should be clear, concise, identifiable, and measurable. In this regard, the following elaboration applies:

Specific: What is entailed in the innovation? What is it? What does it do?
Measurable: Can the accomplishments be quantified? Can it be tracked? Is documentation available?
Achievable: Is it realizable? Is it repeatable? Is it sustainable over the long haul?
Relevant: Is the innovation related to a mission? Is it needed in the context of organizational performance?
Timely: Is the innovation appropriate for the current period or scenario? Is it noticeable within a bounded interval of time? Does it have a definite beginning and a definite end?

SYSTEMS-BASED METRICS DEVELOPMENT

When we talk of innovation, we often focus only on the technological output of the effort. But more often than not, innovation is predicated on the soft side of the enterprise, including people and process. The technical side of innovation cannot happen unless the people and process sides are adequately included. Thus, a systems theoretic approach is essential to realizing the goals of innovation. Everything about innovation is predicated on a systems view of the mission environment. A system is often defined as a collection of interrelated elements whose collective output is higher than the sum of the individual outputs. For the purpose of innovation, a system is a group of objects joined together by some regular interaction or interdependence toward accomplishing some purpose. An innovation system must be delineated in terms of a system boundary and the system environment. This means that all organizational assets impinge upon the overall output. The resources applied to the progression of innovation consist of three organizational assets, namely:

1. People
2. Technology
3. Process

A fourth resource is emerging that is possibly unfamiliar to the reader but alas, must disrupt the comfort of having only three factors to concern oneself with when designing a systems model:

4. Data analytics

Data analytics is becoming so pervasive in itself that soon (if not already) it will *not be possible* to innovate without having imbued the organization with a considerable framework and methodologies (in other words, strategy) for collecting, sanitizing, analyzing, and reviewing data. Since effective data analytics will serve as a canvas for each of the three prior resources, it should be called out on its own, until it becomes as natural as air. Since it is not yet as natural as air, modern data analytics, utilizing the benefits of Big Data, Cloud Computing, and novel software, will be called out in this book, especially when its application has particular benefit for innovation methods.

The efficient and effective application of these assets is what generates the desired output of innovation. The quantitative methodology presented in this book focuses on the people aspect of innovation. Specifically, we consider the learning curve implication (as a metric) of people in an innovation environment. Related quantitative methodologies can be developed for technology management and process development.

An innovation system may be a team or organization, consisting of many elements that interrelate and interact with one another and with the environment within which the system operates. The health and wellbeing of the whole system depends on the health and wellbeing of all the interrelated and interacting elements (particularly people) and the effectiveness of its responsiveness to the challenges in its environment. In the context of innovation, a system involves the interactions between people, technology, and process, and the effects of modern data analytics. Systems thinking is a mindset, which applies the systems approach to analyze and synthesize an organization's operations with the objective of resolving system deficiencies. The core of systems thinking rests in the ability to discern patterns that adequately describe the organization and the people within it. Consultations with various experts (Brezinski, 2018; Barlow, 2018; Rusnock, 2018; Baker, 2018; McCauley, 2018) yielded a summary of additional considerations for the pursuit of a successful innovation program. Rusnock (2018) provided insights into typical questions related to agility and innovation in a complex jet-fighter technology acquisition environment.

- What is management's experience with the Agile Principles? What are the priorities? Are the priorities known and accepted by everyone?
- What is the experience with having all the right individuals in the program for innovative improvements?
- Are the correct metrics in place for assessing the outputs of innovation? Have they changed?
- What innovative methodologies are being employed and where?

UMBRELLA THEORY FOR INNOVATION

The umbrella theory for innovation was introduced by Badiru (2019). In this chapter, that theory is adapted as a template for assessing the scope and hope for innovation. Hope is warranted for uncovering nooks and corners of the spectrum of an organization, where innovation can take root and thrive. An extensive literature review concludes that an overarching theory was lacking to guide the process of innovation. The key to a successful actualization of innovation centers on how people work and behave in team collaborations. Hence, the methodology of the "umbrella theory for innovation" takes into account the interplay between people, tools, and process. The template presented here is illustrated in Figure 2.1. The umbrella theory for innovation capitalizes on the trifecta of human factors, process design, and technology tool availability within the innovation environment. The template harnesses the proven efficacies of existing tools and principles of systems engineering and management. Two specific options for this purpose are the Triple C principles of project management (Badiru, 2008) and the DEJI Systems Model (2014, 2019, 2023) shown in Figure 2.2.

No innovation effort can be successful without taking care of the human aspects that drive an organization toward its goals. The mental or cognitive capabilities of each person will determine how he or she responds to collaborative opportunities in any innovation environment. A manifestation of Maslow's Hierarchy of Needs can guide each person's responsiveness and adaptation to innovation opportunities. A team member at the lowest level of the hierarchy will respond differently from someone at the higher levels of the hierarchy of needs. Concurrent with the individual hierarchy of needs is the organizational hierarchy of needs, which is where

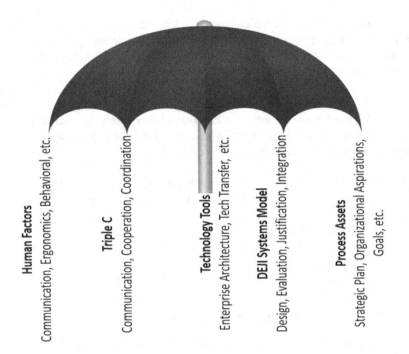

FIGURE 2.1   Umbrella Template for Innovation Metrics

innovation pursuits normally originate. A multidimensional representation of the dual hierarchy of needs forces a consideration of a wider scope of human and organizational needs in the pursuit of innovation.

Communication at each level in the hierarchy must be customized for that level. The Triple C principle of project communication (Badiru, 2008) provides the framework for innovation communication, cooperation, and coordination, as described in Chapter 9. For innovation integration, the DEJI Systems Model is deemed applicable for innovation: Design, Evaluation, Justification, and Integration. This is covered in Chapter 8 of this book.

A semantic network, also called a frame network, is a knowledge base that represents semantic relationships between elements in an operational network or system. It is often used for knowledge representation purposes in software systems. In innovation, a semantic network can be used to represent the relationships among elements (people, technology, and process) in the innovation system. This representation can give a visual cue to the critical paths in the innovation network. From Badiru (2023) and affiliated innovation diffusion researchers (Brezinski, 2018), five qualities are postulated as the requirements for the success of an innovation effort:

1) **Relative advantage**: This is the degree to which an innovation is perceived as better than the idea it supersedes by a particular group of users, measured in terms that matter to those users, like economic advantage, social

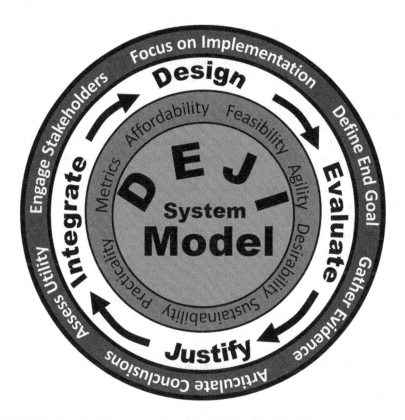

FIGURE 2.2 Graphical Representation of the DEJI Systems Model

prestige, convenience, or satisfaction. The greater the perceived relative advantage of an innovation, the more rapid its rate of adoption is likely to be. There are no absolute rules for what constitutes "relative advantage." It depends on the particular perceptions and needs of the user group.

2) **Compatibility with existing values and practices**: This is the degree to which an innovation is perceived as being consistent with the values, past experiences, and needs of potential adopters. An idea that is incompatible with their values, norms, or practices will not be adopted as rapidly as an innovation that is compatible.

3) **Simplicity and ease of use**: This is the degree to which an innovation is perceived as difficult to understand and use. New ideas that are simpler to understand are adopted more rapidly than innovations that require the adopter to develop new skills and understandings.

4) **Trialability**: This is the degree to which an innovation can be experimented with on a limited basis. An innovation that is triable represents less uncertainty to the individual who is considering it.

5) **Observable results**: The easier it is for individuals to see the results of an innovation, the more likely they are to adopt it. Visible results lower uncertainty and also stimulate peer discussion of a new idea, as friends and neighbors of an adopter often request information about it.

INNOVATION READINESS MEASURE

Table 2.3 illustrates a framework for an innovation assessment tool. The layout is to assess the readiness of an organization on the basis of desired requirements with respect to pertinent factors.

Based on the spread of innovation requirements over the relevant factors in Table 2.3, a quantitative measure of the innovation readiness of the organization can be formulated as follows.

Assuming that each checkmark can be rated on a scale of 0 to 10, the following composite measure can be derived:

$$IR = \sum_{i=1}^{N} \sum_{j=1}^{M} r_{ij}$$

where:

IR = Innovation Readiness of the organization
N = Number of Requirements
M = Number of Factors
r_{ij} = alignment measure of requirement i with respect to factor j

This measure can be normalized on a scale of 0 to 100, on the basis of which organizations and/or units within an organization can be compared and assessed for innovation readiness. Obviously, an organization that is competent in executing and actualizing innovation will yield a higher innovation readiness measure.

DIVERSITY OF INNOVATION

With reference to the proverbial summons and sermon of striking while the iron is hot, the iron of innovation is very hot right now. The aim of this book is to strike while the "innovation iron" is hot, prevalent, and pervasive in business, industry, academia, government, and the military. To this end, metrics, rubrics, and standards for innovation are needed.

In response to the prevailing priority of business and industry to drive innovation, this chapter introduces the prevalence and diversity of innovation in how people work and collaborate in the pursuit of personal and organizational goals.

The focus of innovation ranges from the acquisition of technology products to human services and organizational processes as well as workforce talent development. This justifies using a systems framework for the methodology presented in this book.

TABLE 2.3
Tabulation of Innovation Assessment Tool

Innovation Requirements	Mapping of Organizational Assets to Drive Innovation						
	Factor 1: Technical Assets	Factor 2: Organizational Human Resources	Factor 3: Financial/Cost Spectrum	Factor 4: Mission Priority	Factor 5: Defense Priority	Factor 6: Process Integrity	Factor 7: Tech Transfer
Ability to facilitate technology	✔	✔	✔			✔	
Integration of disparate technical groups	✔						
Ability to resolve conflicts			✔		✔	✔	
Intellectual alignment with national needs				✔			✔
Ability to deflect service loads of staff	✔	✔	✔	✔	✔		
Ability to respond to market concerns		✔				✔	✔
Ability to integrate vertically with organizational goal						✔	

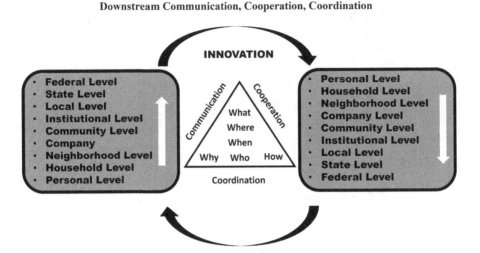

FIGURE 2.3   Bi-directional Flow of Triple C Communication for Innovation Information

Every organization in business, industry, and government is professing to be pursuing innovation. Human needs as well as organizational needs typically drive innovation.

The pressing questions about innovation center around the following:

- What exactly is innovation?
- Who caters to innovation?
- Where is innovation applicable?
- When do we know we have innovation?
- Why is innovation essential?
- How do we assess and sustain innovation?

These questions necessitate the broad perspectives of systems thinking toward metrics, rubrics, and standards for innovation. Structured communication is essential for conveying the needs and requirements of innovation both at the individual level and at the organizational level. For this purpose, we leverage the Triple C model (Badiru, 2008), from the principles of project management, for the framework of driving innovation. The bi-directional flow of information in an innovation environment is depicted in Figure 2.3. The idea of the process is that innovation information flows vertically in both directions as well as horizontally in both directions.

Badiru (2019) introduced the umbrella theory for innovation. Practical applications of this inclusive model of innovation are presented in Badiru (2020), Badiru and Lamont (2022), Badiru and Barlow (2019), and Badiru (2014). As a discussion example, the national defense acquisition system is an excellent platform for innovation analysis because it encompasses various aspects of engineering, science, and technology. For system benchmarking discussion purposes, the defense acquisition system is the management process by which the US Department of Defense provides effective, affordable, and timely systems to entities within the Department of Defense. Requirements, funding, and acquisition process form the core of the defense acquisition system. As a specific focus, the methodology of this book uses a systems-oriented approach to develop techniques and strategies for driving innovation. Of particular interest is the view of the innovation ecosystem as a learning system. A learning system is a sustainable system that feeds off leveraging new knowledge and operational processes. The national goal of achieving a progressive and sustainable acquisition can be advanced through systems theoretic methodologies. The methodology advocated here centers on how people communicate, cooperate, coordinate, and collaborate to actualize the concepts and ideas embedded in innovation initiatives. What does it mean to pursue and actualize innovation? In the concept of this book, the answer is a mix of quantitative and qualitative processes that facilitate innovation metrics, rubrics, and standards in any organization.

The theme of innovation is presently sweeping around the world. Organizations in business, industry, government, academia, and the military are all embracing innovation with different flavors of conceptual and practical pursuits. The process of managing and actualizing innovation can be ambiguous and intractable, because innovation is not a tangible product that can be assessed with traditional performance metrics. From a control perspective, innovation is nothing more than using a rigorous management process to link concepts and ideas to some desired output. That output will be in the form of one of three possibilities:

- Product (a physical output of innovation)
- Service (a provision resulting from innovation)
- Result (a desired outcome of innovation)

Essentially, innovation is the pathway from an initial conceptual point to a discernible end point, as illustrated in Figure 2.4. The ingredients of innovation are the following:

1. Technology framework, upon which innovation is expected to happen
2. Workforce, upon whose education, training, and experience innovation is supposed to rest
3. Operational process, on the basis of which actions take place to make innovation happen

Successful innovation is predicated on a solid foundation and interplay of people, technology, and process. In the hypothetical rendition of Figure 2.4, the dependent variable is "innovation output." The independent variable is the aggregated resource level, which is composed of people, technology, and process. Those resource elements could, themselves, be dependent on other organizational assets, thereby

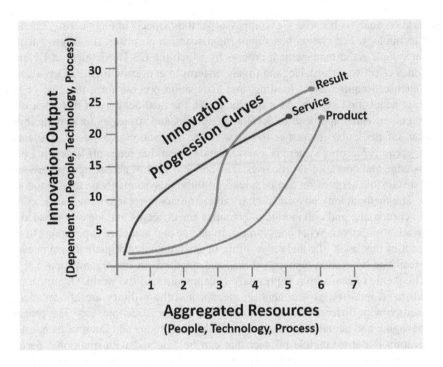

FIGURE 2.4   Innovation Progression Curves with Respect to Desired Outputs

creating hierarchical embedment of interrelated elements in a systems structure. The end point of each curve in the figure represents either a desired result, an expected service, or a required product.

Innovation is multidimensional and has been addressed from different perspectives in the literature. Keeley et al. (2013) discuss ten types of innovation, as enumerated here:

1. Profit Model
2. Network
3. Structure
4. Process
5. Product Performance
6. Product System
7. Service
8. Channel
9. Brand
10. Customer Engagement

Innovation types 1 through 4 are categorized as falling under the group heading of "Configuration." Types 5 and 6 are grouped under "Offering." Innovation types 7 through 10 fall under "Experience." Configuration-based innovation is focused on the inherent workings of an enterprise and its business system. Offering-based innovation is focused on the enterprise's core product or service, or a collection of its products and services. Experience-based innovation is focused on more customer-directed elements of an enterprise and its business system. The authors emphasize that this categorization does not imply process timeline, sequencing, or hierarchy among the types of innovation. In fact, any combination of types can be present in any pursuit of innovation. Thus, the framework embraced by any innovation-centric organization can be anchored and initiated at any of the ten types. This free-flow concept fits the innovation systems premise in this book. Viewed as a system, the pursuit of innovation can include a variety of elements, working together, to produce better overall output for the organization.

Voehl et al. (2019) present a collection of topics that can make up the Innovation Body of Knowledge (IBOK). The concept, tools, and techniques presented in their book reinforce the need to take a systems view of innovation. What is covered in the book includes preparing the organization for innovation, promoting and communicating innovation, creativity for entrepreneurship both personal and corporate, innovation process model, business readiness for innovation, building organizational foundation for innovation, interdisciplinary approach to TRIZ (Theory of Inventive Problem Solving) and STEM (Science, Technology, Engineering, and Mathematics), and intellectual property management for innovation. The diversity of topics and other literature sources confirm that innovation is not just one "thing." A systems thinking approach is, indeed, required to analyze and synthesize all the factors involved in innovation.

A broad, intensive, and detailed review of the literature on innovation confirms the multifaceted nuances and requirements for driving innovation in a defense acquisition system. Badiru (2023) presents several case examples of where and how innovation is desired in the US Military Strength. Satell (2017) presents what he calls a playbook for navigating a disruptive age for the purpose of following a mapping scheme through the latest technological developments. Hamel (2012) covers what matters in innovation pursuits in terms of values, passion, adaptability, and ideology in an innovation environment. Degraff and Degraff (2017) highlight the essentiality of constructive conflict in the pursuit of innovation. Schilling (2018) uses a story-telling approach to highlight traits, foibles, and ingenuity in breakthrough innovations. Personalities profiled in the book include Albert Einstein, Elon Musk, Nikola Tesla, Marie Curie, Thomas Edison, and Steve Jobs. Lockwood and Papke (2018) cover the deliberate pathways for accomplishing innovation through personal dedication. Mehta (2017) uses the "biome," a large naturally occurring community of flora and fauna occupying a major habitat (e.g., forest or tundra), to illustrate how a fertile environment can be created to facilitate a natural occurrence of innovation. Essentially, his hypothesis is the creation of a business environment where innovation can occur and thrive. Verganti (2009) highlights the importance of design in facilitating competition, which drives innovation. A cultural comparative study of military innovation in Russia, the US, and Israel is the focus of Adamsky (2010). He studied the effect of different strategic cultures on the approaches to military innovation in the three countries. Lessons learned from each culture can influence innovation responses in the other countries. There is an art to innovation, as opined by Kelley (2016). Most of the case examples described in the book point to the need to consider human factors and ergonomics in the pursuit of sustainable innovation. On the flip side of the art are the myths of innovation humorously described by Berkun (2010). Grissom et al. (2016) provide six pieces of evidence of innovation in the US Air Force. The literature review confirms that innovation is not new to the US Air Force and has been practiced since the official birthday of the US Air Force on September 18, 1947. Humans have pursued and actualized innovation for centuries. What is different now that suddenly makes innovation a hot topic in today's operational climate? The conjecture is that the increasingly complex and global interfaces of our current civilization call for new ways of doing things. Thus, innovation is simply a new way of doing things. As far back as 1985, Drucker (1985, 2013) pointed out the discipline of innovation. Even farther back, in 1962, Rogers (1962) presented the incipient theory of the diffusion of innovations. Things that have been assumed and done by default in the past now require new looks and innovative approaches. That means that innovation has "the need for change" as its causal foundation. Consider the air travel security–centric changes that have occurred since September 11, 2001 (aka 9/11). The changes are innovative and responsive to the threats of today. Anyone or any organization that is ready and receptive to change is essentially embracing, practicing, and actualizing innovation, which will lead to achieving the benefits of innovation. On this basis, the methodology of this book centers on modeling a change-focused environment to facilitate innovation.

RISK MATRIX FOR INNOVATION METRICS

Not every innovation produces the desired or advertised results. There are risks, and there are potentials. As part of the interest in assessing innovation, both must be rigorously evaluated. A matrix layout facilitates looking at the multidimensionality of innovation. Leveraging the integration requirement of the DEJI Systems Model, we envision research to develop some sort of a risk matrix for innovation metrics by listing the likelihood versus risk of impact of new innovation across the spectrum of expected products. In this regard, risk and likelihood are measured over the categories of low, medium, and high. In this way, the likelihood versus potential impacts could be assessed. Along these lines, Badiru (2023) presents Badiru's Systems Integration Framework, which considers People, Policy, and Technology in the innovation environment (see Figure 2.5).

CONCLUSIONS

This chapter presents a broad coverage of the definitions, the dimensions, the prevalence, and the diversity of innovation in many things we do. Innovation provides the foundation for the advancement of society by significantly changing the way we live. The entire spectrum of life is affected, from the standpoint of the simplest of consumer products to the portfolio of national needs. The needs and priorities that drive innovation are described within the context of a multidimensional rendition of

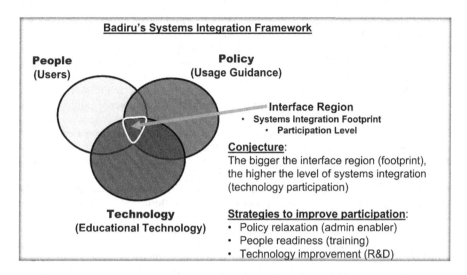

FIGURE 2.5   Badiru's Systems Integration Framework

Maslow's Hierarchy of Needs. The essential role of communication is highlighted as the lead focus in the Triple C model to facilitate innovation, cooperation, and coordination. Innovation progression curves are presented with respect to the desired outputs in any operational system.

REFERENCES

Adamsky, D. (2010). *The Culture of Military Innovation: The Impact of Cultural Factors on the Revolution in Military Affairs in Russia, the US, and Israel*, Stanford, CA: Stanford University Press.

Badiru, A. B. (2023). *Systems Engineering Using DEJI Systems Model: Design, Evaluation, Justification, and Integration with Case Studies and Applications*, Taylor & Francis CRC Press, Boca Raton, FL.

Badiru, A. B. (2008). *Triple C Model of Project Management: Communication, Cooperation, and Coordination*, CRC Press, Boca Raton, FL.

Badiru, A. B. (2014). Quality Insights: The DEJI Model for Quality Design, Evaluation, Justification, and Integration. *International Journal of Quality Engineering and Technology*, Vol. 4, No. 4, pp. 369–378.

Badiru, A. B. (2019). Quality insight: Umbrella Theory for Innovation: A Systems Framework for Quality Engineering and Technology. *International Journal of Quality Engineering and Technology*, Vol. 7, No. 4, pp. 331–345.

Badiru, A. B. (2020). *Innovation: A Systems Approach*, Taylor & Francis CRC Press, Boca Raton, FL.

Badiru, Adedeji B. and Cassie B. Barlow, Eds. (2019). *Defense Innovation Handbook: Guidelines, Strategies, and Techniques*, Taylor & Francis/CRC Press, Boca Raton, FL.

Badiru, Adedeji B. and Gary Lamont (2022). *Innovation Fundamentals: Quantitative and Qualitative Techniques*, Taylor & Francis CRC Press, Boca Raton, FL.

Baker, Bud (2018, November 29). *Strategy and Technology: The First and Second Offsets*, SENG 599 Seminar Presentation, Air Force Institute of Technology, Dayton, OH.

Barlow, Cassie (2018, October 11). *Innovative Mentoring for Organizational Success*, SENG 599 Seminar Presentation, Air Force Institute of Technology, Dayton, OH.

Berkun, S. (2010). *The Myths of Innovation*, O'Reilly Media, Inc, Sebastopol, CA.

Brezinski, Brad (2018, November 8). *Strategy and Innovation*, SENG 599 Seminar Presentation, Air Force Institute of Technology, Dayton, OH.

Bottiau, D. (2018, January 13). *Iterative Innovation vs Disruptive Innovation*. Medium. Retrieved May 14, 2023, from https://medium.com/@dbottiau/iterative-innovation-vs -disruptive-innovation

Degraff, J. and S. Degraff (2017). *The Innovation Code: The Creative Power of Constructive Conflict*, Berrett-Koehler Publishers, Inc, Oakland, CA.

Drucker, P. F. (1985). *Innovation and Entrepreneurship: Practice and Principles*, Harper & Row Publishers, Inc, New York.

Drucker, P. F. (2013). The Discipline of Innovation. In Harvard Business Review. *HBR's 10 Must Reads on Innovation*, Harvard Business Review, Boston, MA.

Grissom, A. R., L. Caitlin, and K. P. Mueller (2016). *Innovation in the United States Air Force: Evidence from Six Cases*. Santa Monica, CA: RAND Corporation. https://www .rand.org/pubs/research_reports/RR1207.html

Hamel, G. (2012). *What Matters Now: How to Win in a World of Relentless Change, Ferocious Competition, and Unstoppable Innovation*, 1st edition. Wiley Professional Development (P&T).

Keeley, L., R. Pikkel, B. Quinn, and H. Walters (2013). *Ten Types of Innovation: The Discipline of Building Breakthroughs*, Wiley, Hoboken, NJ.

Kelley, T. (2016). *The Art of Innovation*, Profile Books Ltd, London.

Lockwood, T. and E. Papke (2018). *Innovation by Design: How Any Organization Can Leverage Design Thinking to Produce Change, Drive New Ideas, and Deliver Meaningful Solutions*, Career Press, Newburyport, MA; RAND Corporation, Santa Monica, CA.

Marchet, Gino, Marco Melacini, Chiara Sassi, and Elena Tappia (2017). Assessing Efficiency and Innovation in the 3PL Industry: An Empirical Analysis. *International Journal of Logistics Research and Applications*, Vol. 20, No. 1, pp. 53–72 https://doi.org/10.1080 /13675567.2016.1226789.

McCauley, Pamela (2018, September 27–29). *Industrial Engineering: The Education for Innovators*, Keynote Address, 2018 Industrial Engineering and Operations Management Conference, Washington, DC.

McClory, Eileen (2023). Sinclair Upgrading Drone Operator Program with Federal Award. *Dayton Daily News*, April 22, 2023, page B4.

Mehta, K. (2017). *The Innovation Biome*, River Grove Books, Austin, TX.

Rogers, E. M. (1962). *Diffusion of Innovations*. The Free Press (Macmillan), New York.

Rusnock, Christina (2018). *Innovation and Technology Capability Pipeline*, SENG 599 Seminar Presentation, Air Force Institute of Technology, Dayton, OH, November 15, 2018.

Satell, G. (2017). *Mapping Innovation: A Playbook for Navigating a Disruptive Age*, McGraw-Hill, New York.

Schilling, M. A. (2018). *Quirky: The Remarkable Story of the Traits, Foibles, and Genius of Breakthrough Innovators Who Changed the World*, Public Affairs, Hachette Book Group, New York.

Verganti, R. (2009). *Design-Driven Innovation: Changing the Rules of Competition by Radically Innovating What Things Mean*, Harvard Business School Publishing, Boston, MA.

Voehl, F., H. J. Harrington, R. Fernandez, and B. Trusko (2019). *The Framework for Innovation: A Guide to the Body of Innovation Knowledge*, CRC Press, Boca Raton, FL.

3 Concept of Standards, Rubrics, and Metrics for Innovation

INTRODUCTION

There is a broad tapestry of innovation pursuits floating around in business, industry, academia, government, and the military. Unless there is some sort of standards, rubrics, metrics, and/or benchmarks, we will all be spinning the wheels of innovation in whichever way we can. In this chapter, we present general advocacy and guidance for the pursuit of innovation. Both qualitative and quantitative approaches are addressed. Standards, whether agreed upon, moderated, or even concocted, can provide assessment templates and yardsticks for assessing where innovation is happening and how to advance and sustain it.

What Is a Standard?

A standard is a template that provides rules, guidelines, characteristics, and/or expectations for accomplishing a specific goal. A standard, typically archived as a document, can also be provided verbally or numerically. A standard can be used to ensure that materials, products, processes, and services meet specified levels of performance with respect to safety, reliability, durability, maintainability, reproducibility, and quality. A standard is usually used to measure, calibrate, and/or reaffirm compliance. A standard is normally established by consensus and approved by a recognized authority, such as a regulatory body or an industry association. In the case of innovation metrics, a generally recognized authority is presently lacking, although individual groups may subscribe to existing standards pertaining to specific industry or commerce. The theme of this book may help in pushing for an innovation-themed standard, even if it is in localized operating settings.

What Is a Rubric?

A rubric is typically an evaluative tool that sets the expectations for expected outcomes. It lists criteria and the associated levels of expected quality for each criterion. A rubric is essentially a scoring scale. We can look at it as a scale or ruler for assigning performance levels. Rubrics are most often used in learning, training,

DOI: 10.1201/9781003403548-3

or knowledge-transfer settings. Since innovation is very generic, each organization will have to determine what rubrics are appropriate, effective, and realizable for the innovation products of interest. A question of interest here is how to assign scores to innovation levels when "innovation" itself is nebulous. Our argument here is that the thought of a rubric may push "innovation pushers" to think about the value and worth of what is being touted as "innovation." There is no mathematical formula in this regard to translate a claim of innovation on some linear scale to the value that the innovation purports to have.

What Is a Metric?

A metric is a quantifiable measure that helps to track and evaluate a product, service, or result. A metric can be used to describe or measure a specific characteristic or attribute of a product. With a metric, the path to accomplishing the advertised goals of innovation can be ascertained. Quantifiability, in this case, does not have to be on a numeric scale. As presented in Chapter 1, alternate measurement scales suitable for this purpose could be nominal or ordinal. For example, how do we know if one innovation in aviation is better than another innovation in the same industry? The conjecture presented by the **"Badiru-Tourangeau Matrix of Innovation Metrics (BT-MIM)"** may offer some insights for readers, decision-makers, policy-makers, proponents, and advocates.

What Is a Benchmark?

A benchmark is a yardstick (or norm) used to compare performance between multiple entities, either animate or inanimate. In addition to being an inter-entity comparative tool, a benchmark can also be used as an absolute comparison against an accepted standard. In this regard, how do we benchmark two aviation innovation products from two competing aviation organizations? Again, this book is not advocating an iron-clad norm for innovation but rather, a sensitivity to what is expected out of innovation, albeit with differing foci from different organizations.

Methodology of BT-MIM

Based on all the preceding definitional narrative, we now propose the Badiru-Tourangeau Matrix of Innovation Metrics (BT-MIM) to guide organizations toward being conscious and sensitive to what innovation entails. What we present here is the shell of the methodology, which may spark interest in further research, development, and implementation of the methodology. The framework of the methodology is simply a matrix of innovation attributes against rating levels of performance. Table 3.1 shows the outline structure of BT-MIM.

The functional form of BT-MIM is represented generically as shown here:

$$I = f(x,y,z)$$

TABLE 3.1

BT-MIM: Badiru-Tourangeau Matrix of Innovation Metrics

| Innovation | | | | Measurement Scale | |
Attributes	High	Medium	Low	(Quantitative, Qualitative)	N/A
Cost					
Schedule					
Quality					
Safety					
Relevance					
Attribute abc					
Attribute def					
Attribute ghi					

where the functional form is subject to the specific operating environment of interest and the attendant objectives and constraints, where x, y, and z are generic terms representing variables of interest. This is presented generically as fodder for further research.

CONCEPT OF STANDARDS FOR INNOVATION

Standardization can facilitate project coordination and organizational learning. The growing number of international projects and widespread adoption of project management are leading to a significant increase in the number of individuals across the world who need to communicate within and understand the field of project management.

Standards provide a common basis for global commerce. Without standards, product compatibility, customer satisfaction, and production efficiency cannot be achieved. Just as quality cannot be achieved overnight, compliance with standards cannot be accomplished instantaneously. The process must be developed and incorporated into regular operating procedures over a period of time. Standards define the critical elements that must be taken into consideration to produce a high-quality product. Each organization must then develop the best strategy to address the elements. Standards can be formulated under three possible avenues:

1. Regulatory standards
2. Industry consensus standards
3. Contractual standards

All three types of standards are essential for developing widely applicable project management standards. Regulatory standards refer to standards that are imposed by a governing body, such as a government agency. All firms within the jurisdiction of the agency are required to comply with the prevailing regulatory standards. Consensus standards refer to a general and mutual agreement between companies to abide by a set of self-imposed standards. Contractual standards are imposed by the customer based on case-by-case, order-by-order, or project-by-project needs. Most international standards will fall in the category of consensus. Lack of international agreement often leads to trade barriers by nations, industries, and special interest groups.

An international standard for terminology and concepts would prevent many misunderstandings and increase efficiency of project management. Just as in the case of quality standards, unified standards for project management will aid consistent understanding of expectations. Standardization of project management processes under the guidelines of ISO (International Standards Organization) will clarify and unify industry-to-industry practices. ISO standards are accepted across the world and carry a level of authority that is recognized beyond those involved with project management professional associations. It is on the basis of this realization that ISO standards for project management are being pursued by cooperating professional organizations. Once those standards become available, project management will reach new implementation heights and be more widely acclaimed.

INNOVATION COORDINATION

Based on whatever standards, rubrics, and metrics are developed and applicable, innovation coordination will be essential to accomplish the expected outcomes of innovation. We present this section within the context of viewing innovation as a project pursuit, thereby following the coordination framework of conventional project management (Badiru, 2008). Coordination is the key that really gets things done in an innovation project after the initial period of project trepidation (worrying and anxiety) has been transcended through "thinking and planning." No project system operates in isolation. Each project must interact within and externally to its scope of operations. There will be interactions with multiple subsystems across different organizational platforms with respect to the pursuit and actualization of innovation. This calls for coordination at various stages of a project of innovation.

As an anecdotal example, Highway Safety advocates often cite a lack of coordination among state motor vehicle bureaus as a big reason why high-risk traffic offenders fall through the cracks and commit interstate infractions again and again. There is often good communication and cooperation among the bureaus. But without the coordination phase to actualize a cooperative agreement, nothing gets done effectively, regardless of whatever innovation is being claimed across state lines. If any public safety undertaking is approached from a project management perspective, under the structure of Triple C (Badiru, 2008), more effective operations can be achieved.

Innovation's effectiveness should not be measured by mere volumes of resources allocated to it. Success, in a contemporary project, is predicated on effective coordination of human resources, work processes, and tools. So, what is coordination? Coordination refers to working with other people, allies and/or competitors. For example, collecting information requires coordinating with other people who own, protect, or manage the information. Frequently, we talk of synergy of a work team. Innovation synergy does require a high degree of coordination across cooperating platforms. Project coordination can be defined as follows:

> **Innovation coordination is a balanced choreography of teamwork across the various elements of an innovative organization and among several members of the innovation project team.**

Each team member must exhibit conscientious commitment to the project so that project harmony can be assured. If project communication is done properly as the first stage of the Triple C approach, commitment of team members becomes more realizable.

INITIATING COORDINATION

Project coordination requires mutual understanding of the goal and schedule of the project. That mutual understanding emanates from proper Triple C communication and cooperation. A primary requirement for a project manager is to constructively influence workers during the coordination phase of a project. Coordination doesn't just happen. It must be effected through direct actions under the leadership of a capable project manager. Proclamations and memos are effective in initiating coordination points in a project.

There needs to be a call-to-order aspect of project coordination after the phases of communication and cooperation have been accomplished. Coordination must be an iterative process. Each stage of coordination initiates the next coordination stage. For coordination to be sustainable, from a metrics standpoint, "bureaucratic acrobatics" and "administrative gymnastics" must be avoided. Bureaucratic acrobatics refers to unnecessary bureaucracy that encumbers the work process through pretentious activities that feign high skills and performance excellence. For example, pontification and long-winded styles of operation are symptoms of a process grounded in bureaucracy. Project bureaucrats tend to act in self-important ways, especially when not qualified to handle a process efficiently. Verbal acrobatics is the process of saying a lot without really saying much, thereby impeding coordination toward project progress. Similarly, administrative gymnastics refers to the process of dancing around the core issues in a project. Focusing on tangential issues, rather than the "meat" of a project issue, impedes project coordination. Perambulatory maneuvering of tasks defeats the purpose of project coordination. Innovation project teams must be watchful and wary of verbal acrobatics and verbal gymnastics and not be swayed by one person's personality. Cohesive organization of efforts is required to achieve project coordination. Important elements of coordination in an innovation environment include:

- Balancing of tasks
- Validation of time estimates
- Authentication of lines of responsibility
- Identification of knowledge transfer points
- Standardization of work packages
- Integration of project phases
- Minimization of change orders
- Mitigation of adverse impacts of interruptions
- Avoidance of work duplication
- Identification of team interfaces
- Verification of work rates
- Validation of requirements
- Identification and implementation of process improvement opportunities

ADAPTIVE PROJECT COORDINATION

Selecting appropriate and adaptive organizational structures for project coordination is essential for project success. Coordination must be effected across both the managerial and technical processes of a project, as illustrated in Figure 3.1. In the figure, project coordination is modeled as a function of managerial process, technical process, adaptive project planning, and adaptive organizing. Coordination is the common thread that links all the requirements. A managerial process should be continually reviewed, modified, and improved to adapt to current business needs. The concept of continuous process improvement (CPI) facilitates an adaptive managerial process. Adaptive project planning infers using contingency planning to respond to prevailing developments in the project environment. No plan should be cast in concrete. A plan should be developed to have avenues for modifications as new realities of the project are encountered. Adaptive organizing refers to the ability of the existing project structure to assume new physical forms and operational constructs based on contemporary needs of the organization. The technical process of an organization

FIGURE 3.1   Innovation Coordination across Managerial and Technical Processes

is the core asset that transforms concept into product. As new technologies develop, it is through an adaptive technical process that an organization can take advantage of the technologies. All of these require careful coordination, in deed rather than just rhetoric.

COORDINATION FOR INNOVATION SCHEDULING

Project scheduling is the most visible part of project management because a schedule indicates the beginning and end points of a project. A schedule is effected under the Triple Constraints of cost, time, and performance expectations. The Triple C approach links the Triple Constraints affecting a project to the Triple C processes of communication, cooperation, and coordination to ensure project results. Figure 3.2 shows the top-down and bottom-up information flows between the Triple Constraints side and the Triple C side in a project system. Cost, schedule, and performance expectations must be linked to the communication, cooperation, and coordination requirements. Project cost is affected by team members' reactions to the communication, cooperation, and coordination aspects of a project. In the same way, performance is influenced by levels of communication, cooperation, and coordination. Finally, all these factors collectively determine how well a project schedule is formulated, executed, and sustained.

CPI requires coordination across teams. Figure 3.3 presents a graphic depiction of team interfaces for improvement efforts. Continuous coordination involves passing the "baton" from one team to another team and from one project stage to the next. Coordination interfaces are both internal as well as external.

There is a dilemma with the conventional fluctuating approach to performance improvement. In this case, the process starts with a certain level of performance. A

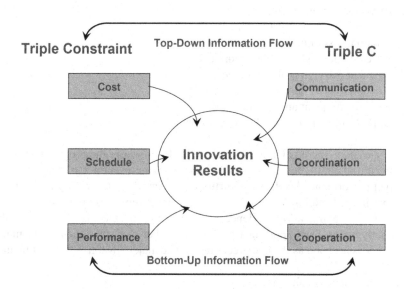

FIGURE 3.2   Coordination Linkages to the Triple Constraints

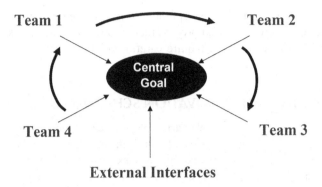

FIGURE 3.3   Innovation Team Interfaces for Coordination

certain performance level is specified as the target to be achieved by time T, which is the end of the planning horizon. Without proper coordination and control, the performance will gradually degrade until it falls below the lower control limit at time $t1$. At that time, a drastic effort will be needed to raise the performance level. If neglected once again, the performance will go through another gradual decline until it again falls below the lower control limit at time $t2$. Again, a costly drastic effort will be needed to improve the performance. This cycle of degradation–innovation may be repeated several times before time T is reached, at which time a final attempt will be needed to suddenly raise the performance to the target level. But unfortunately, it may be too late or too expensive to achieve the target performance. There are many disadvantages of the conventional fluctuating approach to improvement. They are:

1. High cost of implementation
2. Need for drastic control actions
3. Potential loss of project support
4. Adverse effect on personnel morale
5. Frequent disruption of the project
6. Too much focus on short-term benefits
7. Need for frequent and strict monitoring
8. Opportunity cost during the degradation phase

As an alternative, there is the approach of CPI. In this case, the process starts with the same initial quality level, and it is continuously improved in a gradual pursuit of the target performance level. As opportunities to improve occur, they are immediately coordinated and implemented. The rate of improvement is not necessarily constant over the planning horizon. Hence, the path of improvement is curvilinear rather than strictly linear. The important aspect of CPI is that each subsequent performance level is at least as good as the one preceding it. The major advantages of CPI include:

1. Better client satisfaction
2. Clear expression of project expectations
3. Consistent pace with available technology

4. Lower cost of achieving project objectives
5. Conducive environment for personnel involvement
6. Dedication to higher-quality products and services

A concept similar to CPI is continuous measurable improvement (CMI). Continuous measurable improvement is a process through which employees are given the authority to determine how best their jobs can be performed, measured, and coordinated. Since the employees are continually in contact with the job, they have the best view of the performance of the process. The employees can identify the most reliable criteria for measuring the improvements achieved in the project. Under CMI, employees are directly involved in designing the job functions. For example, instead of just bringing in external experts to design a new production line, CMI requires that management get the people (employees) who are going to be using the line involved in the design process. This provides valuable employee insights into the design mechanism and paves the way for the success of the design as a customer-centric design.

INNOVATION COST CONTROL

As a graphical approach, we can develop a plot of innovation cost versus time for projected cost and actual cost. The plot permits a quick identification of the points at which cost overruns occur in the pursuit of innovation. In order to close the cost–benefit gaps, coordination across functions must occur. We consider a case of periodic monitoring of project progress. Cost is monitored and recorded on a periodic basis (e.g., monthly). If cost is monitored on a more frequent basis (e.g., days), then we may be able to have a more rigid control structure. Of course, the decision-maker or analyst will need to decide whether the additional time needed for frequent monitoring is justified by the extra level of control provided. The control limits may be calculated with the same procedures used for X-bar and R charts in quality-control programs, or they may be based on customized project requirements. In addition to drawing control charts for cost, we can also draw control charts for other measures of performance such as task duration, quality, or resource utilization. Several aspects of a project can contribute to the overall cost of the project. These aspects must be carefully tracked during the project to determine when control actions may be needed. Some of the important cost aspects of a project are:

- Cost estimation approach
- Cost accounting practices
- Project cash flow management
- Company cash flow status
- Direct labor costing
- Overhead rate costing
- Incentives, penalties, and bonuses
- Overtime payments

The process of controlling project cost covers several key issues that management must coordinate throughout the organization. These include:

1. Proper planning of the project to justify the basis for cost elements
2. Reliable estimation of time, resources, and cost
3. Clear communication of project requirements, constraints, and available resources
4. Sustained cooperation of project personnel
5. Good coordination of project functions
6. Consistent policy for project expenditures
7. Timely tracking and reporting of time, materials, and labor transactions
8. Periodic review of project progress
9. Revision of project schedule to adapt to prevailing project scenarios
10. Evaluation of budget depletion versus project progress

These items must be evaluated as an integrated control effort rather than as individual functions. The interactions between the various actions needed may be so unpredictable that the success achieved on one side may be masked by failure on another side. Such uncoordinated analysis makes cost control very difficult. The project managers must be alert and persistent in the cost coordination and monitoring function. Some government agencies have developed cost control techniques aimed at managing large projects that are typical in government contracts. The cost and schedule control system (C/SCS) is based on WBS (work breakdown structure), and it can quantitatively measure project performance at a particular point in a project. Another useful cost control technique is the Accomplishment Cost Procedure (ACP). This is a simple approach for relating resources allocated to actual work accomplished. It presents costs based on scheduled accomplishments rather than as a function of time. In order to determine the progress of an individual effort with respect to cost, the cost/progress relationship in the project plan is compared with the cost/progress relationship actually achieved. The major aspect of the ACP technique is that it is not biased against high costs. It gives proper credit to high costs as long as comparable project progress is maintained.

INNOVATION INFORMATION FLOW

As the complexity of systems increases, the information requirements increase. Innovation management is essential because it offers a systematic approach to information exchange for enterprise-wide coordination. Innovation projects, in particular, require a well-coordinated communication system that can quickly share the status of each element of the innovation pursuit. Reports on individual elements must be tied together in a logical manner to facilitate managerial situational awareness and control. The innovation coordinator must have prompt access to individual activity status as well as the status of the overall project. A critical aspect of this function is the prevailing level of communication, cooperation, and coordination. The project management information system (PMIS) has evolved as the solution to the problem of monitoring, organizing, storing, and disseminating project information. Many commercial software products have been developed for this purpose. The basic reporting elements in a PMIS may include the following:

- Financial reports
- Project deliverables
- Current project plan
- Project progress reports
- Material supply schedule
- Client delivery schedule
- Subcontract work and schedule
- Project conference schedule and records
- Graphical project schedule (Gantt Chart)
- Performance requirements evaluation plots
- Time performance plots (plan versus actual)
- Cost performance plots (expected versus actual)

Many standard forms have been developed to facilitate the reporting process in general project management. These could be equally useful for innovation management and control. Coordination across project elements requires that everyone be on the same page. This requires communication. A responsibility matrix should be used to clarify project requirements. The matrix can help resolve questions such as the following:

- Who is to do what?
- How long will it take?
- Who is to inform whom of what?
- Whose approval is needed for what?
- Who is responsible for which results?
- What personnel interfaces are required?
- What support is needed from whom and when?

INNOVATION INTEGRATION ACROSS DISCIPLINES

Innovation management is a comprehensive endeavor that covers diverse knowledge areas. In addition to the primary body of knowledge defined by the Project Management Institute in the PMBOK™, there are several elements of operation that complement the obvious areas. There are several subtle areas that are just as important. The view here encompasses all those areas that are important to embrace in light of the emerging complexity and global aspects of digital-based innovation projects. While no single innovation manager or innovation coordinator is expected to be well versed in the expanded body of knowledge of innovation, it is important to be situationally aware of the elements and to know where to get interfacing help when dealing with those elements. A taxonomy of selected disciplinary elements is presented in Table 3.2.

There are several dimensions to achieving an expanded body of knowledge for innovation project management purposes. The dimensions range from issues that are customer centric and market driven to those that are organizationally focused. An innovation system must continually stay on top of most, if not all, of the dimensions by pursuing the guidelines outlined here.

TABLE 3.2

Taxonomy of Innovation Integration across Disciplines

Qualitative Elements	Quantitative Elements
Technology Assessment	Systems Engineering
Marketing	Operations Research
Organizational Behavior	Financial Analysis
Entrepreneurship	Economic Analysis
Intrapreneurship	Decision Analysis
Group Learning	Cognitive Science
Business Enterprise	Productivity Analysis
Knowledge/Technology Transfer	Probability & Statistics
User Interface	Forecasting
Innovation Management	Supply Network Analysis
Strategic Planning	Systems Simulation

Accountability: Ensuring that everyone knows exactly what they must do in order to meet project goals.

Benchmarking: Conducting programs that objectively compare the organization's project performance with that of competitors.

Business Direction: Operating with and consistently communicating a clear objective of the project.

Cash Flow Management: Consistently having the ability to meet short-term and long-term project cash flow obligations.

Competent People: Attracting, continuously developing, and retaining competent and capable employees at all levels of the project organization.

Competitive Pricing: Regularly offering products and/or services at a competitive price within and outside the project boundaries.

Cost Controls: Consistently keeping project spending within budget allocation levels.

Customer Feedback: Actively listening to the customers and using their inputs to improve products and services.

Customer Needs: Regularly asking customers what they want, what can be done for them, and how it can be done better.

Customer Relationship: Developing and supporting products and/or services based on a long-term project partnership rather than short-term benefits.

Customer Responsiveness: Quickly and effectively responding to the needs of project stakeholders.

Customer Satisfaction: Listening carefully to customers and addressing any issues that may cause disaffection.

Customer Service: Consistently providing service and support to ensure that the customer is satisfied with products and services.

Decision-making: Making effective decisions on a timely basis within the constraints of time, cost, and performance expectations.

Development Systems: Operating with systems that guide the development of products and services from the initial ideas to actual offering in the marketplace.

Ethical Behavior: Having integrity and being responsible in choosing morally correct actions to achieve goals while understanding the differences and similarities between ethical standards and personal ethics.

Financial Stability: Continuously having adequate financial resources to sustain organizational growth or stability.

Integration of Work Processes: Linking operational and functional relationship of key work processes within a project system.

Life Cycle Management: Skillfully managing products and services as they pass through the product life cycle stages (introduction, growth, maturity, and decline) in accordance with the iterative steps of project management.

Market Image: Communicating a clear and consistent message highlighting the benefits associated with project outputs, whether in terms of physical products or services.

Marketing Plan: Developing and using a marketing plan that will achieve the objectives of the project.

Measurements: Consistently using accepted performance standards to measure and assess project results.

Motivation: Enrolling and mobilizing all employees to support the project's vision of success.

New Product Development: Steadily developing and producing new products and services that offer additional value from the project.

Operating Plans: Operating with well-coordinated project plans that are revised and updated as necessary.

Operational Performance: Operating with work processes to produce products and services at the cost, quality, and schedules required by project stakeholders.

Operational Systems: Developing operational systems that allow the organization to grow in an effective manner.

Organizational Structure: Operating with a flexible organization that is specifically designed for accomplishing tasks.

Performance Reviews: Conducting performance reviews on a formal and regular basis and clearly communicating project status.

Plan of Action: Setting goals and executing action plans that ensure the project will stay on a positive course.

Product and Service Enhancements: Proactively and deliberately improving the characteristics and benefits of all products and services on a continual basis.

Product and Service Quality: Consistently striving to achieve 100% quality and reliability in all products and services.

Product Testing: Actively obtaining reactions and feedback about products and services from project customers and stakeholders.

Productivity Improvement: Striving to increase gains from physical and human resources through increased output relative to input.

Profitability: Maximizing financial performance despite the need for other investments such as new product development, training, new facilities, and other infrastructure.

Quality Management: Uniformly operating with an all-encompassing philosophy of management based on a vision of quality and customer satisfaction.

Reliable Forecasts: Accurately estimating how much of the organization's goods and services must be produced to meet future project needs.

Rewards and Recognition: Rewarding people for finding solutions; providing positive recognition for the accomplishment of goals.

Risk Taking: Encouraging and taking calculated risks in setting and reaching goals to increase the rate of growth, sales, profits, and competitive position.

Sales Leadership: Developing and maintaining a well-trained, highly motivated sales force to disseminate project information promptly.

Sales Leads: Regularly developing high-quality sales opportunities as avenues to maximize project acceptance.

Sales Organization: Operating with systems that support the sales staff in developing new customers and maintaining existing accounts.

Shareholder Value: Continually building increased monetary value for shareholders and project stakeholders and protecting their investments on an ongoing basis.

Skill Development: Continuously training employees to develop and maintain competitive skills that will advance the project toward the eventual goal.

Teamwork: Consciously building and using teams of individuals to achieve results, identify new opportunities, and solve problems.

Vision: Consistently communicating what the project organization represents or strives to achieve.

Work Environment: Creating a work environment that is conducive for the project team to work and utilize opportunities for achievement, growth, and accountability.

OODA LOOP APPLICATION TO INNOVATION

Figure 3.4 shows an OODA (Observe, Orient, Decide, Act) loop application within the context of Triple C. The centerpiece of the application is a gig-saw model of how tasks within an organization must be observed, oriented, decided, and acted upon. Communication, cooperation, and coordination processes help the organization to achieve the intended end results.

BUILDING A VALUE CASE FOR INNOVATION

Any project can benefit from a process of building value and increasing performance. Project management touches every aspect of an organization. It can, thus, be instrumental in building an innovation value and performance for the organization. Total systems integration facilitates a value stream across the organization. Innovation value is defined as the level of worth, importance, utility, or significance

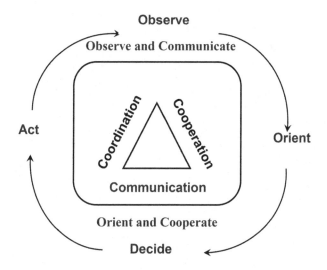

FIGURE 3.4 OODA Loop Application to Triple C

associated with a project with an organization (see BT-MIM tabular template). An **Input-Value-Output** process transfers inputs (e.g., raw materials, technology, personnel) to desired outputs (e.g., products and services). Typically, value translates to profit for the organization. The merit and justification for an innovation project is a function of several factors, as depicted in Figure 3.5. For instance, sales and service provides a linkage to the customer through product follow-ups, repairs, and upgrades. R&D provides the basis for creating, procuring, and implementing new technology for productive application within a project. Process planning, particularly in manufacturing, refers to the design of detailed work processes. Manufacturing involves creating value for marketing through an organization's production facilities. Quality assurance refers to the process of achieving and sustaining preservation of product value as perceived by the customer. Maintenance (Mtce) involves keeping the productive infrastructure of an organization in peak performance form. Purchase orders transpire through an organization through supply chain channels to provide an organization with products and services needed to generate the organization's outputs. Accounts and finance adds value to the organization by ensuring a balance of credits and debits in the organization's resources. Training closes the loop of input-value-output by ensuring that personnel have clear job objectives and fully understand the what-why-who-how-when aspects of the organization's operations. A proper integration of all these factors helps to uphold a consistent value stream for a project.

Innovation value stream mapping (VSM) is an important technique used to identify points of value-added contribution to an innovation project. VSM is normally done using manual drawing and flow-through analysis of the tasks and steps that constitute a project operation. In manufacturing, VSM is done on the shop floor using a series of sticky notes on a wall chart. As has been reported in the literature, hand-drawn value maps are not very effective communication tools. An analyst can

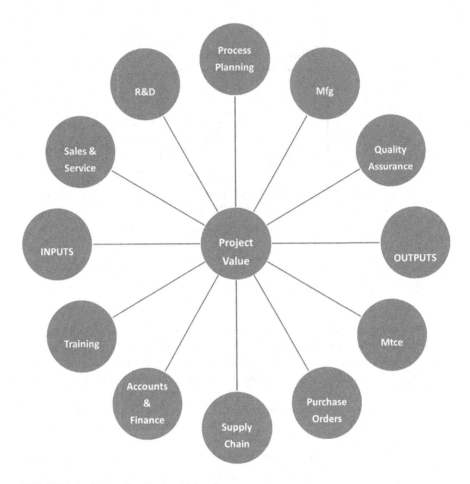

FIGURE 3.5   Innovation Project Value Components

enhance the VSM charts by incorporating simple communication steps into the process. This requires asking each participant to think through his or her respective operations. We should recall that Triple C requires full and early communication of why particular tasks are needed. In the same way, a full and early explanation of why a process needs to be mapped can enhance the VSM technique.

The details collected by the analyst are transcribed into presentation graphics that can be studied as a simulation model of the actual process. A primary purpose of doing value stream analysis is to identify areas of process waste in accordance with lean principles. Waste is generally defined as any activity that does not create value for the end product. The first step in the waste-cleanup process is to take an inventory of all the activities making up a process, with the goal of identifying those that produce value and those that do not. An important tool for this inventory is the value stream map. A value stream map is a representational diagram that describes the sequence of activities a business undertakes to produce a product or a family of similar products.

Process steps that are identified as not adding value to the overall project effort are eliminated, thereby conserving limited resources for the more value-adding steps. "Lean" means the identification and elimination of sources of *waste* in operations. The basic principle of "lean" is to take a close look at the elemental compositions of a process so that non-value-adding elements can be located and eliminated.

VSM describes the activities required to produce a family of products from the beginning to the end. The process involves walking through the process and literally following the product through each step it undergoes. The VSM analyst records every process encountered using very specific symbols that are used to represent the flow of material and information through the system. The result is a current state map. The symbols help to quickly identify processes that add value and those that do not. Using the current state map as a baseline, the analyst can then develop future state maps: that is, the ideal system designs that reduce or eliminate the non-value-added steps. The key to using VSM effectively is its *symbology* and *iconography*. While VSM was developed primarily for manufacturing and related processes, it can be adapted for project value analysis to track the flow of materials and information through the project life cycle. The communication component of Triple C, in particular, can enhance VSM. Conversely, the process of documenting VSM findings can enhance communication processes in a project.

KAIZEN ANALYSIS OF AN INNOVATION PROJECT

By applying the Japanese concept of *Kaizen*, which means "take apart and make better," an organization can redesign its processes to be lean and devoid of excesses. In a mechanical design sense, this can be likened to finite element analysis, which identifies how the component parts of a mechanical system fit together. It is by identifying these basic elements that improvement opportunities can be easily and quickly recognized. It should be recalled that the process of WBS in project management facilitates the identification of task-level components of an endeavor. Consequently, using a project management approach facilitates the achievement of the objectives of "lean."

ORGANIZATIONAL LEARNING FOR INNOVATION

Organizations learn and advance as a collective body. The Triple C model can facilitate organizational interactions that enhance group learning and advancement. A Triple C application to organizational coordination and learning has the following elements:

- Diagnose the work culture and social climate of the organization
- Assess employees' readiness for an organizational restructuring to meet project goals
- Develop a strategic communication plan
- Prepare the employees for new challenges brought on by the project

We can develop a web plot of factor relations for organizational learning. The central factor in that plot will be the primary organizational goal, which is supported by the ambient factors within the organization. Such factors may involve the different business units of the organization working synergistically together in pursuit of the central goal. There is cross-pollination of ideas between the factors. The angular direction between the factors indicates a measure of factor relationship, proximity, or mutual relevance. The plot can be drawn to increase or decrease the space between two factors to indicate the degree of factor closeness. If one factor fails, a disruption to the flow of work may develop; but some learning effect may still occur from the failure. This creates an experience base that serves the organization better in subsequent endeavors.

The following summary presents the major steps of project management within the context of organizational advancement from project initiation through project phase-out. It should be noted that each step provides a learning opportunity for the organization. The consistent platform for feed-forwarding the learning from each step to the next is the sequence of communication, cooperation, and coordination that exist in the project. The phase-out of the project itself represents a learning opportunity that is transferred to the next project endeavor. The control across the phases of innovation consists of the following steps:

1. Innovation initiation
2. Innovation communication
3. Innovation cooperation
4. Innovation coordination
5. Innovation phase-out

INNOVATION ORGANIZATIONAL LEADERSHIP BY EXAMPLES

Leadership of innovation is not unlike conventional leadership of organizational pursuits with respect to the expected standards, rubrics, and metrics. There follow some pertinent ideas:

- Leadership System that defines and communicates organizational direction, vision and major objectives
- Strong Leadership team that is not dependent on just one individual
- Leadership that sets goals to improve performance in areas of health and environment protection
- Leadership that sets examples of being good corporate citizens
- Leadership that designs job and organization structures to promote empowerment, efficiency, employee development, and elimination of non-value-added efforts
- Leadership that empowers employees and teams to implement suggestions rather than relying on suggestion systems
- Leadership that eliminates functional departments and layers of management where possible

- Senior project leadership involvement in promoting customer focus and performance excellence
- Leadership involvement in promoting customer focus and performance excellence toward how an organization contributes to the community as a corporate citizen
- Leadership approach to recognition programs tailored to individual and group preferences

INNOVATION LEARNING THROUGH PARTICIPATION

The following are some pertinent questions to ask in ensuring that employees learn through participation in innovation project affairs.

- What is done to ensure the workforce is focused and engaged in satisfying customer expectations?
- What is done to encourage employee participation in project and process improvement?
- How do you motivate the workforce to participate in total project management?
- How does an organization ensure proper resources are made available to employees so that they can participate in total project management?
- How are employee contributions recognized and rewarded?
- How is teamwork encouraged throughout the project system?
- How does the organization communicate the effects of future changes to the workforce?

CONCLUSIONS

In summary, innovation project coordination offers the avenue through which organizational efforts are brought into fruition (Badiru, 2019). As a concluding statement, the primary lesson of the Triple C model presented in this chapter is not to take cooperation for granted, for the sake of standards, rubrics, and metrics of innovation. Cooperation must be pursued, solicited, secured, and preserved explicitly. The process of securing cooperation requires a structured communication upfront. It is only after cooperation is in effect that all innovation efforts can be coordinated.

REFERENCES

Badiru, Adedeji B. (2008). *Triple C Model of Project Management: Communication, Cooperation, and Coordination*, Taylor & Francis CRC Press, Boca Raton, FL.
Badiru, Adedeji B. (2019). *Project Management: Systems, Principles, and Applications*, 2nd edition, Taylor & Francis CRC Press, Boca Raton, FL.

4 Statistical Engineering Approach to Innovation Metrics

INTRODUCTION

The benefit of statistics for project control must not be understated. With evolving technology, the methods and techniques to manage data are changing rapidly, almost as rapidly as data is being generated. Thankfully, the math is the math, and basic statistical tenets remain as they are: normal distributions, standard deviation, the law of averages, and data patterns have almost a universal quality to them that will go unchanged, despite the technological maturation around them. Learning ways to program them for data analysis becomes an art that the reader can adopt to become a master of project control. Mastering project control is a fundamental skill of industrial engineers. By following the teachings in this chapter, the reader can reap the benefits of project control mastery without needing a degree in Industrial and Systems Engineering (ISE). Fundamental principles of using ISE methodologies of statistical inferences can be found in Badiru et al. (2012), Badiru and Kovach (2012), Badiru (1993), and Badiru (2014).

Innovative knowledge development processes are essential to support Department of Defense (DoD) complex weapon systems development, a process characterized by rapidly maturing technologies. Unfortunately, there is ample evidence suggesting that the current system lacks such approaches. The implementation of an innovative knowledge development process requires a culture change beyond only relying on engineering judgement to one that is information driven. This process of change is achieved through policies, workforce development, executive training, and facilitators, such as a center of excellence. This chapter addresses a case study in implementing such an innovative knowledge development process. This process accomplishes its development objectives through close integration with the systems engineering methods. Several components inform systems engineering design decisions, including contractors, program engineers, system subject matter experts, and test results. Through collaboration with the program engineers and subject matter experts, the Scientific Test and Analysis Techniques in Test and Evaluation Center of Excellence at the Air Force Institute of Technology (AFIT) is at the forefront of changing the DoD acquisition culture by injecting rigorous, defensible, and innovative test methodologies and processes into this knowledge accumulation.

DOI: 10.1201/9781003403548-4

Engineering expertise alone used to be sufficient to inform whether a DoD weapon system would perform its tasks and meet requirements. More recently, these weapon systems have many more capabilities and consequently, are extremely complex. With more capability, however, come more subsystems that must meet their own requirements. There is then a need to understand the performance, reliability, integration, and interactions of these subsystems early in the weapon system's development before large corrective costs manifest. The complexity arising from these integrated and interacting subsystems and larger systems can no longer be adequately informed by engineering expertise alone. To make informed decisions on these increasingly complex systems, a culture shift within DoD acquisitions must occur so that it moves from a culture reliant on engineering judgement to one that is information based.

Typical DoD weapon system development transitions through four phases: capabilities identification, technology development, systems technology integration for engineering and manufacturing development, and production. The capabilities identification phase conducts analysis to identify what future capabilities are needed and determine specific measurable requirements of the potential system. The technology development phase identifies the technology that currently exists and its level of maturity, and facilitates rapid advancement of any technology that is needed to achieve the system requirements. The systems technology integration phase for engineering and manufacturing development focuses on the systems engineering (SE) task of integrating technology and re-engineering any shortfalls. Finally, the production phase begins manufacturing the system *en masse*. It is at this point that any system requirement not met or any technology integration issue not successfully addressed typically becomes prohibitively expensive to correct.

Moving from one phase to another requires decisions that are dependent on assessments. These assessments typically rely on subject matter expert judgement as well as a process known as test and evaluation. While subject matter assessments alone are not adequate, when coupled with efficient and effective testing and the corresponding quantitative analysis, a powerful means of knowledge development will result. This knowledge more effectively informs both SE and acquisition decisions. In this chapter, we discuss the inherent complexity of weapon systems, provide a brief history of knowledge development within the DoD, discuss innovation of the defense acquisition program through culture change, highlight the importance of scientific test and analysis techniques (STAT) to develop the foundation of the culture change, and finally, provide the future direction of innovating defense acquisition.

COMPLEXITY OF DOD WEAPON SYSTEMS

Before discussing methods to drive innovative practice into the defense acquisition process, we first explain what makes modern weapon systems so complex. These systems are inherently complex because of their systems of systems (SoS) nature, their reliance in development on modeling and simulation (M&S), their reliance on software, net-centricity, and in the future, their ability to act more autonomously. All of these aspects make efficient and effective testing and assessment of these systems challenging. In the following subsections, we expand on each of these components, all of which are current challenges within defense acquisition.

SYSTEMS OF SYSTEMS ARCHITECTURE

DoD SoS engineering is the design of systems that satisfy specific requirements and is performed under uncertainty of advancing technology and integration of component systems. It focuses on choosing the right systems and their interactions to satisfy requirements in complex environments. In DoD and elsewhere, SoS can take different forms. Based on a recognized taxonomy of SoS, the four types of SoS that are found in the DoD today are virtual, collaborative, acknowledged, and directed (Maier, 1998; Dahmann, 2008). Virtual SoS lack a central management authority and a centrally agreed-upon purpose for the SoS. Collaborative SoS have the component systems interact more or less voluntarily to fulfill agreed-upon central purposes. Acknowledged SoS have recognized objectives, a designated manager, and resources for the SoS; however, the constituent systems retain their independent ownership, objectives, funding, and development and sustainment approaches. Directed SoS are those in which the integrated SoS is built and centrally managed to fulfill specific purposes. Having independent, concurrent management and funding authority at both the component system and SoS levels is a dominant feature of acknowledged SoS. Typically, attention is focused on the management issues that result from the overlapping authority over decisions rather than the technical implications for SE (ODUSD(A&T)SSE, 2008).

There are seven core elements that characterize SE in SoS and which contribute to the complexity of weapon systems. These interconnected elements include: (1) translating SoS capability objectives into SoS requirements, (2) assessing the extent to which these capability objectives are being addressed, and (3) monitoring and assessing the impact of external changes on the SoS. Central to SoS SE is: (4) understanding the systems that contribute to the SoS and their relationships and (5) developing an architecture for the SoS that acts as a persistent framework for (6) evaluating SoS requirements and solution options. Finally, the SoS systems engineer (7) orchestrates enhancements to the SoS, monitoring and integrating changes made in the systems to improve the performance of the SoS. It is this lack of focus on the technical state of the SoS, the metrics that provide knowledge on the developmental state of these elements, and other technical aspects of the SoS that can have significant performance and financial implications during the latter stages of the acquisition process.

MODELING AND SIMULATION

Another source of complexity of weapon systems development is the use of simulations in evolutionary acquisition. Evolutionary acquisition consists of a baseline system being developed and produced with upgrades added at a later date. These upgrades can be either improvements of current capabilities or the addition of new capabilities that were not a part of the initial design requirement. When evolutionary acquisition is pursued, it is often useful to have a validated simulation that possesses appropriate mathematical models to assess performance. "Modeling and simulation (M&S) provides a technical toolset which is regularly used to support systems acquisition and engineering" (ODUSD(A&T)SSE, 2008). M&S is applied throughout the system development life cycle, supporting early concept analysis, design, developmental test and evaluation (DT), integration, and operational test and evaluation (OT).

Because of the characteristics of SoS, M&S can be a particularly valuable tool. Models, when implemented in an integrated analytical framework, can be an effective means of understanding the complex and emergent behavior of systems that interact with each other. Models can provide an environment to help create a new capability from existing systems and consider integration issues that can have a direct effect on the operational user. M&S can support analysis of architecture approaches and alternatives as well as analysis of hardware and software requirements and solution options.

Because it can be difficult or infeasible to completely test and evaluate all the capabilities of a SoS, M&S can be effectively applied to support T&E at different stages in the weapon systems development process. In particular, M&S can be used to understand the end-to-end performance of the overall SoS prior to implementation. In some cases, it is advisable to adopt a model-based process for gaining knowledge of a system. Because of the importance of M&S, it is essential to include planning for M&S early in weapon systems development planning. This planning includes "the resources needed to identify, develop, or evolve and validate M&S to support SE and test & evaluation" (ODUSD(A&T)SSE, 2008). Effectively and efficiently planning M&S into T&E to learn more about the system under development, while a valuable method, adds additional complexity to the current weapon systems.

RELIANCE ON SOFTWARE

Another source of complexity of weapon systems is their reliance on software. Software dependencies within and between systems are also complex, requiring knowledge concerning their performance. "Nearly all modern technology systems depend on software to perform their functions. From remotely piloted aircrafts and smart bombs to self-driving vehicles and advanced fighter jets, software is crucial to the success of today's weapons systems" (IG, 2016).

"The quantity of software that enables weapons systems today drives complexity in engineering, test, and evaluation. Defense systems use hundreds of millions of lines of code generated by defense teams, reused from known government or commercial-off-the-shelf systems, and incorporated from open sources". With this reliance on software, testing systems requires more than engineering expertise to effectively characterize the performance of the system and/or identify shortfalls of the system.

NET-CENTRICITY

Net-centricity is itself an innovative functional approach that requires an innovative knowledge development process. Along with significant increases in software have come increases in networking and information exchanges across countless combinations of system interfaces. Net-centric systems are often characterized as having a service-oriented architecture (SOA), a paradigm for organizing and utilizing distributed capabilities that may be under the control of different owners. A SOA has software architecture where functionality is grouped around processes or capabilities and packaged as interoperable services. A SOA possesses an information technology infrastructure that allows different elements to exchange data or functionality with one another as they participate in the process. The aim is a loose coupling of services and separation of functions into distinct units. This allows accessibility of services over a network in order that they can be combined and reused in the furtherance of mission accomplishment. These services communicate with each other by passing data from one service to another, or by coordinating an activity between two or more services (Dahmann et al., 2009). Incorporating net-centricity into T&E planning of weapon systems is not an easy task because of the interconnected nature of many systems. Henry and Stevens (2009) state that the systems used by all users throughout the DoD are interconnected. These systems include "unmanned aerial systems, handheld systems, ground vehicles, ships, etc." T&E must adapt to requirements of interconnected systems. "The entire enterprise becomes a single complex system comprised of numerous component systems" (Henry and Stevens, 2009).

AUTONOMOUS SYSTEMS

Finally, incorporating autonomy into modern weapon systems is a challenge during T&E. Autonomous systems have gained great interest in recent years and most likely will need to operate in unstructured, dynamic environments. A recent Defense Science Board (2016) study notes that "autonomous systems can be cyber-physical or totally cyber-dominated. In any case, these systems will be dominated by a software architecture and integrated software modules. The DoD historically has had difficulty in specifying, developing, testing, and evaluating software-dominated systems." Particular knowledge development challenges for autonomous systems are noted in a 2016 Scientific Test and Analysis Techniques Center of Excellence (STAT COE) workshop report in the areas of requirements and measures, test infrastructure and personnel, design for test, test adequacy and integration, testing continuum, safety and cybersecurity, testing of human system teaming, and post-acceptance testing (Ahner and Parson, 2016). The current Research & Engineering Autonomy Community of Interest (COI) Test and Evaluation, Verification and Validation (TEVV) Working Group Technology Investment Strategy 2015–2018, signed by the Assistant Secretary of Defense for Research and Engineering (ASD(R&E)), states the need for rigorous test methods of autonomous systems:

Cumulative evidence through RDT&E, DT, & OT – Progressive sequential modeling, simulation, test and evaluation M&S and T&E at each Technical Readiness Level (TRL) and product milestone currently provide an invaluable resource not only to verify and validate that a system satisfies the user requirements, but also to aid in technology development and maturation. However, the development of effective methods to record, aggregate, and reuse T&E results remains an elusive and technically challenging problem.

As just discussed, the complexity of weapon systems is illustrated by their SoS nature, their evolutionary acquisition relying on simulations, their reliance on software, net-centricity, and their ability to act more autonomously. All of these aspects of system complexity make the efficient and effective testing and assessment of these systems challenging. The current acquisition process is heavily reliant on engineering subject matter assessments, which alone are not adequate to make informed decisions. When combined with efficient and effective testing and analysis that generates quality and insightful system performance information, a powerful means of knowledge development results. This knowledge development process that informs DoD complex weapon systems development must support the innovation, agility, and quality of those weapon systems while addressing the aforementioned aspects of system complexity.

BRIEF HISTORY OF TEST AND EVALUATION IN THE DOD AND CURRENT INNOVATIVE EFFORTS

Before presenting a description and method of implementation of this innovative knowledge development process, it is useful to understand the history of knowledge development, namely, T&E, within the DoD. We provide a brief history of T&E and attempts at innovation in the DoD, from which we learn and develop our approach to changing the culture to achieve a system and statistical engineering–enabled approach for process innovation.

In 1971, in order to oversee both DT and OT, the office of Director, Defense Test and Evaluation was formed under the Office of the Director of Defense Research and Engineering, who was responsible for major acquisitions. The office was formed through a series of three memoranda by Deputy Secretary of Defense David Packard in response to recommendations by President Nixon's Blue Ribbon Defense Panel of 1970. In 1977, the need for independent OT saw it moved under the responsibility of the Assistant Secretary Defense for Program Analysis and Evaluation. However, this change lasted only a short time, and in late 1978, it was moved back to the Director, Defense Test and Evaluation. In 1983, independent OT again arose as an issue, resulting in Congress establishing the current office of Director, Operational Test and Evaluation (DOT&E), once again separating the DT and OT functions. In 1994, with the reassignment of live fire testing to DOT&E, the Director Test, Systems Engineering, and Evaluation office was formed. On June 7, 1999 (28 years after Packard created it), Secretary of Defense William Cohen disestablished the test office within what had become the Office of the Undersecretary of Defense for Acquisition and Technology and realigned DT responsibilities as a function under other offices. During those first nearly three decades, all emphasis in T&E in the department continued to be on OT, and Cohen's decision was intended specifically to strengthen the office of the DOT&E; however, it virtually eliminated oversight of DT. Congress would reverse this 10 years later with passage of the 2009 Weapon Systems Acquisition Reform Act, which established the Director, Developmental Test and Evaluation (DDT&E) position (Fox, 2011).

This fluid history of overseeing T&E does not lend itself to supporting innovative acquisition, nor does it lend itself to having a system and statistical engineering–enabled approach for process innovation. However, with the establishment of the DDT&E, which later changed to Deputy Assistant Secretary of Defense for Developmental Test and Evaluation (DASD(DT&E)), DT was about to begin a path toward innovation through the Scientific Test and Analysis Techniques in Test and Evaluation Implementation Plan.

As systems became more complex, testing techniques remained the same. "New programs appear to be more complex than their immediate predecessors in terms of technology, functionality, and, perhaps to a lesser extent, their operational concept" (Drezner, 2009). The relative complexity of the weapon system itself is captured in technical complexity. Elements of technical complexity include the use of electronics, information technology, and software to provide critical functionality and capability beyond more traditional means. That these are increasing can be measured by the percentage of acquisition program funds devoted to these technologies. These technologies reside in sensors, data processing, automation, communication, and data exchange. Many recent weapon systems are multifaceted, multifunction, and multimission systems that include many more specific functions and performance capabilities than predecessor programs (Drezner, 2009).

The proliferation of electronics in both performance and quantity is a major contributor to increasing weapon system complexity (Dietrick, 2006). In directing programs that have been problematic, managers for the government, the prime contractors, and the commercial subcontractors shared one common feature: they underestimated the complexity of requirements, the integration of subsystems, and the interaction of changes in one subsystem with new demands on others (Berteau, 2009).

As the DoD continues to push innovation within its acquisition process, several elements within a framework are required for this innovation (Drezner, 2009):

- National factors, which include education level, strength in science and technology, and supporting infrastructure (e.g., communication and transportation)
- Research & development investment in a wide variety of projects, technologies, and sectors
- Status and attractiveness of the sector (e.g., excitement and dynamism) as indicated by the degree to which industry in that sector is admired by consumers and students, the degree to which it is pushing the state of the art, and its ability to attract and retain top people
- Competition in the sector, as determined by company strategies, industry structure, and rivalry

- Demand conditions – in other words, the customer demanding capabilities requiring innovative new technologies
- Related supporting industries, including lower tiers and science and technology (S&T) base

Additional factors affecting innovation or the conditions that facilitate innovation not explicitly identified in the preceding model include the following (Drezner, 2009):

- An institutional and regulatory environment that encourages new concepts
- Early adopters who are willing to buy and use initial versions of the innovation
- A potential for significant demand for the product
- High potential payoff
- Minimal barriers to entry

INNOVATION OF DEFENSE ACQUISITION PROGRAM KNOWLEDGE DEVELOPMENT THROUGH CULTURE CHANGE

Cultural change is difficult to achieve, especially in large organizations, and the DoD is among the largest in the world. Components of an organization typically consist of purpose and tasks, intellectual or mechanical processes, hierarchy of authority or structure, and people. Within these components is an interlocking set of goals, processes, roles, collaboration, coordination, cooperation, values, attitudes, and assumptions. Over time, these components settle and become a reinforcing system that is difficult to change.

A Forbes article entitled "How do you change an Organizational Culture?" by Steve Denning (2011) presents the strategy that all organizational tools need to be put into play to increase the likelihood of success, but argues that the order matters. These tools consist of leadership tools, management tools, and power tools. Leadership tools entail developing a vision and providing inspiration for change. Management tools include the activities of strategic planning, role definition, incentives, and training. Power tools consist of coercion and regulations. While implementing these tools methodically may lead to cultural changes in some organizations, the DoD is usually considered a bureaucracy characterized by adherence to fixed rules, specialization of functions, and a hierarchy of authority. These characteristics may require an approach to culture change differing from that in smaller organizations. The adherence to fixed rules requires new processes to be addressed by those rules in the form of requirements or regulations. To perform a new function, either a current group must be identified to be trained or educated, or a new specialized group must be resourced and formed. Finally, the new process must be requested or demanded by leadership in authority. To implement an innovative knowledge development process, these challenges need to be overcome to achieve the cultural change desired.

Culture change and innovation are tightly coupled in large organizations. Process innovation intervention is the act of incrementally setting the conditions to achieve enterprise-wide acceptance of an improved process throughout a large organization.

A firm understanding of methods and practices, documented best practices and case studies, and well-written policies and regulations are necessary conditions before culture change can occur within a large organization. This process innovation is depicted in Figure 4.1 as a pyramid, since each element is a foundation of the higher element. As seen in Figure 4.1, a culture change cannot occur without the structure of supportive policies and regulations, which in turn are built around established best practices that highlight the capabilities and strengths of the prescribed methods and processes.

In order to begin the cultural change to improve quantifiable knowledge for acquisition within the DoD, the DASD(DT&E) developed the Scientific Test and Analysis Techniques in Test and Evaluation Implementation Plan in coordination with the Army, Navy, and Air Force T&E executives and the office of the DOT&E. The plan calls for changes in policy, development of the T&E workforce, and establishment of a STAT COE to achieve an innovative knowledge development process with more effective systems and statistical engineering. The STAT COE was established in 2012 to assist acquisition programs in developing rigorous, defensible test strategies as a source for the required high-quality people with advanced degrees and knowledge of the acquisition process. These actions directly contribute to the lower two levels of the pyramid in Figure 4.1 by establishing and implementing STAT in DoD testing.

The DoD's efforts using STAT enable innovation in the form of a more information-based decision-making process in several ways. STAT requires a pool of high-quality people with advanced technical degrees and knowledge of the acquisition process to be effectively implemented. Innovation is achieved incrementally through small continuous improvements in how T&E is conducted. STAT enables technology innovation either by generating the knowledge to mature technologies that fulfill a requirement gap or by generating the knowledge to inform the performance of a

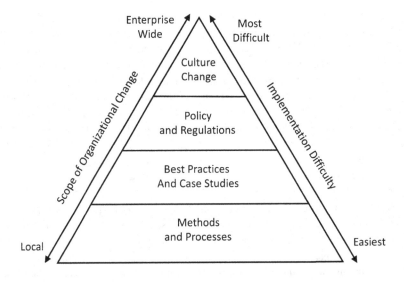

FIGURE 4.1 Process Innovation Intervention

paradigm shift in new technologies, such as autonomous systems, hypersonics, and directed energy. The STAT COE innovates the DoD business model by the creation of a pool of high-quality people who can be drawn upon by acquisition programs but still are considered an integral part of the program and not an outside entity, thus avoiding the problems of additional cost and hiring of scarce human capital. This pool of high-quality people provides a significantly improved service that generates new customer value using (and not replacing) high-quality T&E professionals in other areas. Finally, the STAT COE is innovative in that while adding this capability, it does not offset others in cost within any given acquisition program. The correct use of STAT lowers test costs by more efficiently making use of test resources and lowers acquisition costs through a more effective T&E process, resulting in significantly higher return than the initial cost of the pool of high-quality people.

STAT TECHNIQUES

The STAT strategies allow programs to more effectively quantify and characterize system performance as well as provide information that reduces risk. STAT is defined as the scientific and statistical methods and processes used to enable the development of efficient, rigorous test strategies that will yield defensible results. STAT consists of methods and processes that encompass various techniques including design of experiments (DOE), observational studies, reliability growth, survey design, and statistical analysis used within a larger decision support framework.

The primary challenge in applying STAT to DoD testing is the broad scale and complexity of the systems, missions, and conditions. While advanced STAT methods are used more frequently in industry, manufacturing, or healthcare environments, the DoD lags behind in adopting these techniques, in part because of the complex and constrained nature of DoD testing. In order to address this complex environment, the STAT COE has emphasized a SE approach to decompose the mission, system, or requirement into smaller pieces. One effective way these portions can then be readily translated into rigorous and quantifiable test designs and/or strategies is by using DOE. Figure 4.2 shows a flow diagram that summarizes the infusion of STAT into the DoD T&E process. The procedure begins with the requirement of interest and proceeds through the generation of test objectives, designs, and analysis plans, all of which can be traced directly back to the requirement.

DOE is the systematic integration of well-defined and structured strategies for gathering empirical knowledge about a process or system using statistical methods for planning, designing, executing, and analyzing a test. The end goal of a designed test is to produce clear results and meaningful analysis that leads to informed decisions and the best possible course of action. The DOE process can be explained by following the flow diagram in Figure 4.2.

Within the plan phase of Figure 4.2, there are two actions: understand the requirements and define the design space. The top portion, understand the requirements, is further subdivided into three actions: identify STAT candidates, understand the system and the mission, and determine the test objectives. At this point, there is often a desire to rush to the computer and create a test design, but it cannot happen

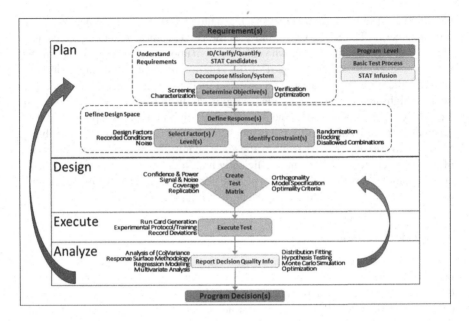

FIGURE 4.2   Schematic of STAT in the Test & Evaluation Process

until T&E personnel fully understand the requirements of the system, the purpose of the test, and how to measure system performance. In other words, creating the test design cannot happen until the test is well planned.

Planning is one of the most difficult phases of the process and cannot take place in a solitary environment. Planning requires input from operators, logistics personnel, analysts, subject matter experts, engineers, and the overall decision-makers to understand the requirements and objectives of the test(s). There are often multiple documents that contain the necessary information, and they can be incomplete and contradictory. Understanding these materials to develop an effective test plan requires a good test team to gather what they can, learn the rest, and (often) make educated assumptions and compromises.

The first step in successfully designing a test is to understand the requirements. Within the DoD, some common reference documents are the test and evaluation master plan and the capabilities development document. Requirements are the starting and end point for T&E. If the requirement is not understood clearly at the beginning of the process, the test team may plan a test that will not produce the data needed to address the requirement and adequately inform the decision-maker. Understanding what is written, what is missing, and/or what needs to be clarified in the requirement is the first step in effective DOE implementation. Without a clear understanding of the requirement, what conditions it pertains to, how it factors into the mission, and how it can and should be tested, the T&E process as outlined in Figure 4.2 is unguided. This crucial first step drives the development of the test objectives, responses, factors, designs, and analysis plans. The amount of detail (or lack thereof) associated with a requirement will directly impact the amount, type, and quality of

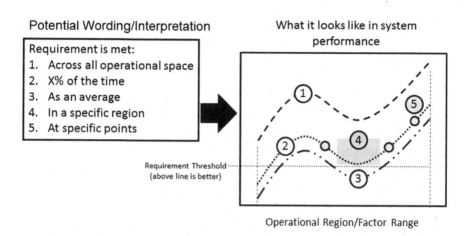

FIGURE 4.3 Translating Requirements to Performance

the data collected from any T&E event. Figure 4.3 depicts a translation of system requirements to performance measures. Key questions to ask when discussing what a requirement says and how it should be evaluated include: 1) What remains to be clarified in the requirement? 2) What is my test objective to address the requirement? 3) Can I effectively characterize the system and if not, where are the un-testable regions? These questions will lead the test team into the design process and help with strategy development and resource planning. The test design must produce data that allows the analysis to address the requirement.

A clear understanding of all requirements by all stakeholders early in the process will inform the planning process, resulting in rigorous and defensible data. Many systems do not meet requirements at the outset, and the program manager desires information on deficiencies so that he can direct resources to correct them. In the SE process, testers help address performance issues and make improvements. Rigorous and defensible data focus this effort.

Once the requirements have been examined and distilled, the next step is to understand the system as it exists and the mission it is designed to perform. From these two steps, the team can define why the test is necessary and clearly define the test goal(s). The goals should be derived from the system requirements and be objective, unbiased, measurable, and of practical consequence (Coleman and Montgomery, 1993). The objectives should be referenced throughout the planning process to ensure that subsequent steps and decisions produce a relevant test plan. Objectives may be to characterize the performance of the system across several test conditions, identify factors that affect the response, validate performance in a simulation, optimize performance of the system in a specific region, or compare new versus legacy systems. If the test has some sort of pass/fail criteria, the consequences of a failure should be noted. They can include rejecting the system, requiring alterations, or simply purchasing an extra unit as a spare.

With the test objectives defined, we can move to the second part of the plan phase in Figure 4.2, and the design space can be created. First, it is important to determine the responses to record that will best address the objectives. The responses are the quantifiable dependent outputs of the system, which are influenced by the independent or controlled variables (factors). The responses must be observable and recordable, and should have a relationship to the test objectives. Defining the response(s) is not always an easy task. The response is ideally a continuous metric as opposed to a binary measure. Many response variables naturally tend to be binary, such as whether a weapon hit or missed a target, a go/no go decision, or whether something is operationally effective or not. However, continuous metrics provide more information than binary metrics and result in more efficiently designed tests, since they require fewer runs. Careful thought should be given to translating binary metrics into continuous metrics. For example, rather than measure hit or miss, measure distance from the aim point. Instead of pass/fail of a quality characteristic crossing a threshold, measure the change in that metric.

After the responses have been established, the next step is to determine the factors, the system inputs that potentially have an effect on the response (and therefore, the performance of the system). There are three types of factors: control, hold constant, and noise. Control factors are purposefully varied during the test so that their specific impact on the response can be measured. Hold constant factors may also have an impact on the response but are held constant and remain unchanged during the test. Hold constant factors may not be of primary interest or may be too difficult to control during the test. Noise factors likely influence the response but cannot be controlled in real life and/or during the test. The noise factors can be further broken down into measurable and unmeasurable noise factors. The measurable factors, such as wind speed or component age, are recorded so that their potential influence on the response can be accounted for. The unmeasurable (and often unknown) factors are best nullified by randomizing the test sequence. Randomization, one of the core principles of DOE, minimizes the effect of lurking variables – those factors that were not accounted for in the test.

A designed experiment focuses on control factors. Typical factors may include configurations, physical and ambient conditions, operator considerations, etc. Factors should not be excluded from consideration without careful thought and analysis. Brainstorming using a cause-and-effect diagram is one of the most effective methods to develop a comprehensive list of potential factors to include in the test. This process must be done collaboratively by the test team so that no factors are unintentionally excluded.

Levels are the values to which each factor is purposefully set during the test. For each factor included in the test, we need to determine the levels that it will be set to. Ideally, the factors are continuous (can take on an infinite number of possible values), since continuous measures provide most information on system performance. Using continuous factors also allows you to make predictions at values of the factors that were not specifically observed in the test. While we gain the most information from a continuous factor, we typically set it to two or three settings initially in a designed experiment. Restricting the number of levels of a continuous factor is done in initial phases of testing because it is easier to identify factor effects. Additional levels can be included in later phases of testing to refine models if necessary, which we discuss in more detail later. The next best option is an ordinal factor, which can be set to a number of fixed ordered values between settings. The final option is a categorical factor, which can have any number of levels with no fixed relationship between them. However, more levels require more test runs to be able to model and determine the impact the factor has on the response.

When all of the factors and levels are agreed upon, there may still be some work to determine whether there are constraints. A factor may be restricted from a level because of limitations to the system or test facility, because of safety, or for any other prudent reason. A disallowed combination occurs when a given set of levels for more than one factor cannot (or should not) be set at the same time. Possible reasons for declaring a set of levels to be a disallowed combination include the inability of the system to operate in that configuration, because the testing facility will not accommodate the configuration, or because there is no value in the information obtained when testing the combination. Examples of these situations are testing a car's cruise control in reverse, attempting a 500-yard shot at a 300-yard shooting range, and testing night vision devices during the day.

One of the primary advantages of using DOE in DoD testing is the ability to build an empirical model of the performance metric. This model allows you to identify the important factors affecting the response in addition to the magnitude and direction of that effect. Well-designed tests also allow you to efficiently identify any interaction effects. Two-factor interactions occur when the effect on the response of one factor depends on the level of another factor. Ahner et al. (2019) present an example of an interaction plot for a notional test with factors range and angle on the response, target location error (TLE). In that example, the difference in TLE due to angle is stronger when range is at the high level. Because the factor levels are actively manipulated by experimenters during the test (rather than simply observed), DOE also allows you to establish causal relationships between the factors and the response.

CREATING A TEST MATRIX

With the plan phase in Figure 4.2 complete, the next step is to create the design matrix. This is done with software that will take into account the difficulty in changing the factors and the purpose of the test. An ideal design is completely randomized to reduce noise, but if some factors are very difficult to change, this may not be feasible. The software will take that into account and randomize where possible. The test matrix is designed in accordance with the test objective, whether that is to screen for important factors, model the response in a specific operating envelope, or optimize the performance of the system.

The choice of the test matrix should be based on the objective of the test itself. A well-designed test will allow you to create a statistical, empirical model to quantify the effects of the factors on the response across the design space. The choice of design will determine the type of model that can be estimated using the results of the test. The model can then be used to predict values of the response throughout the test space, even at conditions not tested. Different objectives will lead to different design choices. For example, if the objective of the test is to identify the factors that have the most impact on the performance measure, a screening design such as a factorial or fractional factorial design is often a good choice. If the objective is to identify the factor levels that optimize a performance measure, a response surface design such as a central composite design may be appropriate. If the test is a computer experiment that has a deterministic response (i.e., you observe the same response value when the same factor levels are input into the experiment), then space filling designs, such as a sphere packing or Latin hypercube design, are common design choices. Computer-generated optimal designs are common design choices when there are constraints on the test space or when the expected model of the response has an unusual form (i.e., there are high-order effects such as cubic terms in the model). Following the process outlined in Figure 4.2 by identifying the test objectives, responses, factors, and any constraints will generally lead to a clear design choice.

When planning a test, it is commonly recommended that at least 80% of the test process should be devoted to the planning phase (Montgomery, 2017). One method to do this is to evaluate a proposed test matrix prior to selecting a final design for the test. One of the biggest constraints in the DoD is a limited test budget, leading to a small number of runs available in the test matrix. The final choice of design must therefore balance the tradeoffs of the run size and the various properties of the design. One metric commonly advocated by T&E personnel is the power of the test, or the probability that a factor effect (main effects and/or two-factor interactions) will be detected given that the effect actually has an impact on the response. The power of the test can be estimated *prior* to testing using common assumptions associated with the empirical model (see Montgomery, 2017 for complete details). Power is dependent on the signal-to-noise ratio (SNR), a ratio of the size of the effect that is practically important to be able to detect divided by the estimated variability due to noise in the system. Power greater than 80% for the desired SNR is a typical threshold when evaluating a design. One of the best (but most expensive) ways to increase the power of the test is to increase the run size.

While power is an important metric used to evaluate a proposed design, there are many other metrics that should be considered when comparing test matrices, including the confounding or alias properties of the design and the prediction variance. The STAT COE is working to build the foundation of the culture change pyramid (Figure 4.1) by advocating the use of these additional metrics when planning a test in DoD acquisitions. We highlight just a few of these methods in the following.

Confounding of effects occurs when a term in the proposed model cannot be distinguished from another term in the model. For example, a two-factor interaction between factors A and B may be confounded with the two-factor interaction between factors C and D. Once the test has been executed, and a model is fit with the data, if the interaction term AB is statistically significant, we cannot resolve whether this effect is actually due to the interaction between AB or between CD. An ideal test matrix will have little to no confounding, so that conclusive decisions can be made after the test. One way to evaluate a design in terms of aliasing is a color map of correlations. Interested readers should refer to Ahner et al. (2019) for further details about the color map example.

An additional design metric that is frequently considered is the prediction variance. A common objective of a test is to make predictions of future values with a

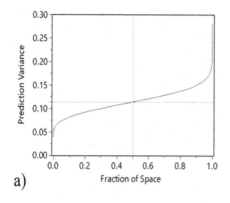
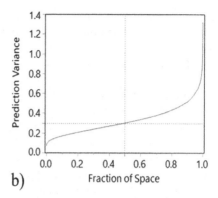

a)

b)

FIGURE 4.4   Example FDS Plots

given degree of confidence. To make these prediction intervals meaningful, the test points should cover the design space, such that the prediction variance throughout the design region is low. One method to analyze the prediction variance is to use a fraction of design space (FDS) plot (Zahran et al., 2003). FDS plots display the prediction variance of the response for a given model across regions of the design space. These plots provide a simple way to compare designs in their potential ability for prediction. Figure 4.4 shows two examples of an FDS plot from the same designs of color maps described earlier. The ideal plot is flat across most of the design region with low values in prediction variance. Note that because of the smaller run size, the prediction variance is much higher for the design in Figure 4.4b.

Another consideration to make when choosing a test matrix is the type of model that can be fit. Screening designs typically allow you to estimate main effects and some two-factor interactions, but not higher-order terms such as quadratic terms. This is because the goal of a screening experiment is to identify the important effects; follow-on testing can be used to refine the model of the response as necessary. If there is previous testing or subject matter expertise that suggests that a higher-order model will be necessary to adequately model the response, the test strategy and choice of design should reflect that knowledge. However, one of the greatest capabilities of using DOE in testing is the ability to test sequentially.

KNOWLEDGE DISCOVERY USING SEQUENTIAL EXPERIMENTATION

As discussed previously, the results of the test can be used to build an empirical model of the response. Many tests initially have a long list of potential factors that may have a statistically significant effect on the response in some way. A common assumption in DOE is the sparsity of effects principle (Montgomery, 2017). Sparsity of effects means that the variability in the response can typically be explained by only a subset of the potential factors in terms of main effects and two-factor interactions. The principle states that higher-order terms are frequently negligible. One large experiment that tests whether every factor has a high-order effect on the response is, therefore, an inefficient test. A better approach is to build the empirical model in stages by testing in phases, so that you start with an initial test that allows you to screen for significant factors and then disregard factors that are not statistically significant in future stages of testing. This process allows you to isolate the key factors that influence the response and refine the empirical model by fitting a potentially higher-order model with the remaining factors.

Figure 4.5 shows a potential progression of testing for a test with four factors. The initial stage allows you to determine the significant factors by focusing on the main effects and some (but not all) two-factor interactions. This initial stage of testing may take ten runs or more if replicates are done. The second phase of testing allows you to resolve any confounded interaction terms among the factors to determine which two-factor interactions are actually causing a change in the response. After the initial two stages of testing, suppose only three factors were determined to be significant. Including center runs (the points in the center of the design space) allows you to determine whether quadratic terms should be included in the model. Stage 3 progresses testing to estimate and determine the statistical significance of the quadratic terms of only the three remaining factors. The final stage typically consists of five to ten test points to validate that the model performs well for conditions in the interior of the design space, typically at locations not previously tested.

In total, the number of runs for this sequential test is between 40 and 45. This efficiency in runs is possible because you can leverage information learned in previous testing to inform the next stage of the test. For example, the initial ranges of factors may be quite wide to identify active main effects of a factor. As testing progresses,

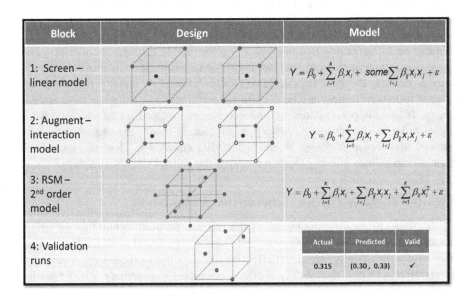

Block	Design	Model
1: Screen – linear model		$Y = \beta_0 + \sum_{i=1}^{k} \beta_i x_i + \text{some} \sum_{i<j} \beta_{ij} x_i x_j + \varepsilon$
2: Augment – interaction model		$Y = \beta_0 + \sum_{i=1}^{k} \beta_i x_i + \sum_{i<j} \beta_{ij} x_i x_j + \varepsilon$
3: RSM – 2nd order model		$Y = \beta_0 + \sum_{i=1}^{k} \beta_i x_i + \sum_{i<j} \beta_{ij} x_i x_j + \sum_{i=1}^{k} \beta_{ii} x_i^2 + \varepsilon$
4: Validation runs		**Actual** 0.315 **Predicted** (0.30, 0.33) **Valid** ✓

FIGURE 4.5 Potential Sequential Test Strategy for a Test with Four Factors (Adapted from Simpson, J., *Testing via Sequential Experiments*, Scientific Test and Analysis Techniques Center of Excellence, Dayton, 2014.)

the range of the levels may decrease in order to focus on a particular region of the design. Alternatively, because there may not be a clear understanding of the factors of a system, the initial ranges of a factor may not be wide enough, and follow-on testing is adjusted to move outside the original test space.

A sequential test strategy allows better knowledge discovery of the system under test. Questions that can be answered more easily using a sequential approach rather than a "one-shot" test include: 1) How do the factors affect the response and by how much? 2) How is the system expected to perform at conditions not tested? 3) Are there areas of the design space that perform better or worse than the specified requirements? 4) Which conditions provide optimal system performance? and 5) What are the tradeoffs in the system if there are multiple, competing objectives? One large test will not be able to answer all of these questions.

EXAMPLE OF SEQUENTIAL TESTING

Consider, for example, a notional test to characterize the performance of a missile warning system. It is unknown how the warning system will behave under a variety of conditions. Responses of interest include the time it takes to notify the presence of a missile and the missile detection rate. There are many factors that may affect the performance of the system as determined by several subject matter experts. These include: the environment (urban vs desert), time of day (morning vs night), target intensity (low watts vs high), approach angle (0 vs 5 degrees), angular motion (0 vs 0.1 rads/sec), sensor resolution (0.1 vs 6 millirads), sensor sensitivity (0.1 vs 5 picowatts/cm^2), and frame rate (30 vs 60 Hz). Due to the sparsity of effects principle, not all of these factors will likely affect the performance of the warning system. A sequential test strategy can first identify the critical few that impact performance. Follow-on testing may then be done to refine the empirical model and perform validation runs.

The subject matter experts initially identified the time of day to be a potential factor and specified two levels: night vs day. As discussed previously, the information obtained from categorical factors is much more limited than that obtained from continuous factors. An alternative measurement of time of day that can be used is illuminance, a measure of the intensity of illumination on a surface. Illuminance, measured in lux, can be as high as 100,000 in direct sunlight and as low as 0.0001 on a moonless, cloudy night. For this test, the illuminance was chosen to range from 5, which represents a dark night, and 10,000, which represents full daylight, but not in direct sun. The target intensity was also initially classified as a categorical factor (low vs high watts). To make this a continuous factor, these levels can be translated into a numeric low and high level using subject matter expertise.

A potential initial test is shown in Table 4.1, with all factors coded to be between −1 and 1. The design is a fractional factorial design that will allow you to determine the main effects and some two-factor interactions that have an effect on the performance of the warning system. The power of the main effects is high (>0.95 for all factors for an SNR of 2), and four center runs are included in the design to determine whether the response is characterized by any quadratic effects. With eight potential

TABLE 4.1
Phase 1 Screening Design for Missile Warning System

Run	Environment	Illuminance	Target Intensity	Approach Angle	Angular Motion	Sensor Resolution	Sensor Sensitivity	Frame Rate
1	Urban	1	1	1	1	-1	-1	-1
2	Desert	-1	1	-1	1	-1	1	-1
3	Urban	-1	1	-1	1	1	-1	0
4	Desert	0	0	0	0	0	0	1
5	Desert	1	1	1	1	1	1	1
6	Desert	-1	-1	-1	-1	1	1	-1
7	Urban	-1	-1	-1	-1	-1	-1	0
8	Desert	0	0	0	0	0	0	-1
9	Urban	1	1	-1	-1	1	1	1
10	Urban	-1	1	1	-1	1	1	1
11	Desert	1	1	-1	-1	1	-1	-1
12	Desert	1	-1	1	-1	1	1	1
13	Urban	0	0	0	0	0	0	0
14	Desert	-1	-1	-1	1	1	-1	1
15	Desert	1	-1	1	1	1	-1	1
16	Desert	1	1	1	-1	1	1	1
17	Urban	1	-1	1	1	-1	-1	1
18	Urban	-1	-1	-1	1	-1	1	-1
19	Urban	1	-1	1	1	-1	1	1
20	Urban	0	0	0	0	0	0	0

factors, this initial screening design can be followed by a second phase of testing once the primary factors driving changes in the response are identified in the preliminary phase 1 analysis.

This example provides the initial phase of a potential sequential experiment. Because there are eight potential factors that may have an effect on the response (in addition to interactions and quadratic terms), designing one large test to investigate all these potential effects is prohibitively large and therefore, extremely inefficient. The current culture in the DoD, however, favors one large test plan. STAT emphasizes sequential learning so that future testing can incorporate the knowledge gained in previous testing. If the initial test in this example indicates that only three factors have a significant effect on the response, the next phases of testing are greatly reduced and focus on the vital few to best refine the empirical model of the performance metrics.

INNOVATION SUCCESS INDICATORS

Looking at success through the Process Innovation Intervention pyramid depicted in Figure 4.1, several key elements have been achieved without achieving the cultural change. Some methods and processes were already well established, such as DOE (Coleman and Montgomery, 1993), while others have been developed, such as that depicted in Figure 4.2, which form a solid foundation of methods and processes. In its first five years, the STAT COE has partnered with over 41 major acquisition programs and has developed and disseminated best practices and case studies to illustrate how to implement these methods and processes within the schedule and budgetary constraints of real-life program management. Acquisition policies and regulations have been changed to require the use of STAT for both DT and operational acceptance testing. The elements in Figure 4.1 appear to have been met without the resulting culture change to a more information-based decision process being fully realized. So, one might wonder, "What is missing?"

In the 2016, DOT&E Annual Report to Congress, the Director stated:

> Since 2012 when the STAT COE was formed, I have noted that programs who engage with the STAT COE early have better structured test programs that will provide valuable information. The STAT COE has provided these programs with direct access to experts in test science methods, which would otherwise have been unavailable.

The Director insightfully noted that the improved outcomes of an information-based decision process that used both engineering judgement and STAT over a mainly engineering judgement–based process occurred for the "programs who engage[d] with the STAT COE early." The initial successes of the STAT COE are encouraging, but many programs still do not have access to personnel that provide the capability of STAT. The STAT COE currently is only able to interact with a subset of major defense acquisition programs, but without additional support, the full potential of the COE cannot be realized. With adequate support from DASD(DT&E) and the program managers, the STAT COE can provide expertise to all levels of acquisition programs and build the base levels of the pyramid in Figure 4.1. This would result in more efficient testing and analysis for the increasingly complex systems throughout the DoD.

The DOT&E recognized this when he stated:

> However, the COE's success has been hampered by unclear funding commitments. The COE must have the ability to provide independent assessments to programs (independent of the program office). Furthermore, the COE needs additional funding to aid program managers in smaller acquisition programs. Smaller programs with limited budgets do not have access to strong statistical help in their test programs and cannot afford to hire a full-time PhD-level statistician to aid their developmental test program; having access to these capabilities in the STAT COE on an as-needed basis is one means to enable these programs to plan and execute more statistically robust developmental tests. Finally, the STAT COE has also developed excellent best practices and case studies for the T&E community.
>
> (DOT&E, 2016)

The DOT&E statement suggests that although the methods and processes are developed, they are only accessible to programs through organizations like the STAT COE; and although best practices and case studies are developed, their accessibility is a challenge and is currently not adequate to achieve the information-based outcomes desired.

In addition to the accessibility of using STAT capability, policy and regulations requiring STAT are necessary, but not wholly sufficient. Without thought leaders who require from their acquisition organizations a transition from an industrial-era engineering judgement decision-making culture to an information-based decision-making culture that uses both STAT and engineering judgement, the gains of using STAT cannot be fully realized. Culture change to a STAT-infused T&E process cannot propagate throughout the DoD without leadership championing the cause. Similarly to the success of Six Sigma in industry (Snee and Hoerl, 2003), unless leadership understands and advocates the use of information-based decision-making through DOE and other statistical techniques, change will not occur.

CONCLUSION

Innovative knowledge development processes are essential to support DoD complex weapon systems development, a process characterized by rapidly maturing technologies. The complexity of these systems arises from their SoS nature, their evolutionary acquisition relying on simulations, their reliance on software, net-centricity, and in the future, their ability to act more autonomously. All of these aspects of system complexity make the efficient and effective testing and assessment of these systems challenging. To meet this challenge requires a transition from an industrial-era engineering judgement–focused decision-making culture to an information-based decision-making culture.

The history of developmental test and evaluation within the DoD is fluid, with a dedicated emphasis on DT within the office of the Secretary of Defense that is inconsistent. With more complex systems becoming the norm, the need for a more information-based decision process that uses both engineering judgement and STAT over the most recent mainly engineering judgement–focused process is required.

The Scientific Test and Analysis Techniques in Test and Evaluation Implementation Plan addresses these changes by incrementally setting the conditions to achieve enterprise-wide acceptance of an improved, innovative knowledge development process throughout the DoD in the form of STAT. This process accomplishes its development objectives through close integration with the SE process. Several components inform SE design decisions, including contractors, program engineers, system subject matter experts, and test results. Through collaboration with the program engineers and subject matter experts, the STAT COE is at the forefront of changing the DoD acquisition culture by injecting rigorous, defensible, and innovative sequential test methodologies and processes into this knowledge accumulation.

STAT are methods and processes that encompass various techniques including DOE, observational studies, reliability growth, survey design, and statistical analysis used with a larger decision support framework. DOE is one STAT technique in particular that has a wide, powerful application. We use this technique to illustrate the STAT innovative knowledge development process.

The initial implementation of STAT has yielded significant results for those programs having STAT expertise available. Engineering expertise alone used to be sufficient to inform whether a DoD weapon system would perform its tasks and meet requirements. More recently, these weapon systems now have many more capabilities and consequently, are extremely complex. There is ample evidence suggesting that the current system lacks such innovative approaches. The implementation of an innovative knowledge development process requires a culture change beyond only relying on engineering expertise judgement to one that is information driven.

To achieve the cultural change desired, both greater accessibility to using STAT capability and thought leaders requiring STAT-developed information are needed. Through collaboration with the program engineers and subject matter experts, the STAT COE is at the forefront of changing the DoD acquisition culture by injecting rigorous, defensible, and innovative test methodologies and processes into this knowledge accumulation.

REFERENCES

Ahner, Darryl, Sarah Burke, and Aaron Ramert (2019). A System and Statistical Engineering Enabled Approach for Process Innovation. In Adedeji Badiru and Cassie Barlow, Eds. *Defense Innovation Handbook: Guidelines, Strategies, and Techniques*, CRC Press, Boca Raton, FL.

Ahner, D. and C. Parson (2016). *Test and Evaluation of Autonomous Systems*, Scientific Test and Analysis Techniques Center of Excellence, Dayton, OH.

Badiru, Adedeji B. (1993). *Quantitative Models for Project Planning, Scheduling, and Control*, Quorum Books, Greenwood Publishing Company, Westport, CT.

Badiru, Adedeji B. Eds. (2014). *Handbook of Industrial & Systems Engineering*, 2nd edition, Taylor & Francis CRC Press, Boca Raton, FL.

Badiru, Adedeji B. and Tina Kovach (2012). *Statistical Techniques for Project Control*, Taylor & Francis CRC Press, Boca Raton, FL.

Badiru, Adedeji B., O. Ibidapo-Obe, and B. J. Ayeni (2012). *Industrial Control Systems: Mathematical and Statistical Models and Techniques*, Taylor & Francis CRC Press, Boca Raton, FL.

Berteau, D. (2009). Forward. In G. Ben-ari and P. A. Chao, Eds. *Organizing for a Complex World: Developing Tomorrow's Defense and Net-Centric Systems*, Center for Strategic & International Studies, Washington, DC, p. ix.

Coleman, D. E. and D. C. Montgomery (1993). A Systematic Approach to Planning for a Designed Industrial Experiment, *Technometrics*, Vol. 35, pp. 1–27.

Dahmann, J. (2008). Systems Engineering for Department of Defense Systems of Systems. In M. Jamshidi, Ed. *System of Systems Engineering*, Wiley, Hoboken, NJ, pp. 218–231.

Dahmann, J., K. J. Bladwin, and G. Rebovich Jr. (2009). System of Systems and Net Centric Enterprise Systems, 7th Annual Conference on Systems Engineering Research, Loughborough.

Defense Science Board. (2016). *Autonomy, Office of the Under Secretary of Defense for Acquisition*, Technology and Logistics, Washington, DC.

Denning, S. (2011, July 23). How Do You Change An Organizational Culture? *Forbes*.

Dietrick, R. A. (2006). Impact of Weapon System Complexity on Systems Acquisition, *Streamlining DOD Acquisition: Balancing Schedule with Complexity*, Center for Strategy and Technology, Air War College, Air University, Montgomery, AL.

Director, Operational Test and Evaluation, (DOT&E) *FY 2016 Annual Report*, Department of Defense, Washington, 2016.

Drezner, J. A. (2009). *Competition and Innovation Under Complexity*, RAND Corporation, Santa Monica.

Fox, J. R. (2011). *Defense Acquisition Reform, 1960–2009, An Elusive Goal*, Center of Military History, United States Army, Washington, DC.

Henry, W. and J. Stevens (2009). Net-Centric System Development, 3rd Annual IEEE International Systems Conference, Vancouver.

Inspector General of the DoD, *DoD Needs to Require Performance of Software Assurance Countermeasures During Major Weapon System Acquisitions*, Department of Defense, Washington, 2016.

Maier, M. W. (1998). Architecting Principles for Systems-of-Systems, *Systems Engineering*, Vol. 1, No. 4, pp. 267–284.

Montgomery, D. C. (2017). *Design and Analysis of Experiments*, 9th edition, Hoboken, NJ: Wiley.

Office of the Deputy Under Secretary of Defense for Acquisition and Technology, *Systems and Software Engineering, Systems Engineering Guide for Systems of Systems, Version 1.0*, Washington, DC: Department of Defense, 2008.

Simpson, J. (2014). *Testing via Sequential Experiments*, Scientific Test and Analysis Techniques Center of Excellence, Dayton, OH.

Snee, R. D. and R. W. Hoerl (2003). *Leading Six Sigma: A Step-by-Step Guide Based on Experience with GE and Other Six Sigma Companies*, FT Prentice Hall, New York, p. 41.

Zahran, A., C. Anderson-Cook , and R. H. Myers (2003). Fraction of Design Space to Assess the Prediction Compatibility of Response Surface Designs. *Journal of Quality Technology*, Vol. 35, pp. 377–386.

5 Engineering Economic Analysis of Innovation Projects

INTRODUCTION FROM ENGINEERING ECONOMIC ANALYSIS

The conventional time value of money is directly applicable to the analysis of innovation from an economic perspective. Recalling the techniques of engineering economic analysis (White et al., 2014; Newnan et al., 2004; Sullivan et al., 2003; Badiru and Omitaomu, 2007; Badiru, 2019), there is a viable path for conducting innovation economic analysis. Financial and economic analyses form the foundation to achieving and sustaining innovation programs. An understanding of the basic techniques of engineering economic analysis can advance how innovation is embraced and leveraged in organizations. The selection of an appropriate problem is extremely important and is a major factor for determining the success of innovation initiatives. A good problem for the pursuit of innovation is one that has the following characteristics:

1. The proposed innovation is affordable
2. The problem affects many people internal and external to the organization
3. There is enough organizational concern about the problem
4. The problem is in a domain where subject matter experts are available
5. Solving the problem has the potential for significant time and cost savings
6. There is a reliable and accessible source of knowledge to be acquired
7. The expected innovation will advance the body of knowledge
8. The innovation is sustainable
9. The innovation can be replicated for organizational or market advancement
10. The innovation can contribute to better efficiency and higher effectiveness

Problem identification refers to the recognition of a situation that constitutes a problem to the organization. Problem identification requires the recognition of a window of opportunity to drive innovation. Both the problem domain and the specific problem must be identified. A problem domain refers to the general functional area in which the problem exists. For example, the general problem of engineering design constitutes a problem domain that design engineers will be interested in. Within engineering design, the specific problem may be that of designing a flexible manufacturing system. The identification of a problem may originate from any of several factors. Some of the factors are:

1. Internal needs and pressures
2. External motivation, such as market competition
3. Management requirement
4. Need for productivity improvement
5. The desire to stay abreast of the technology
6. Technological curiosity
7. Compliance with prevailing rules and regulations
8. Deficiencies in the present process

Once the general problem area has been identified, the next function is to decide what to do about the problem. Several options may be available in addressing the problem. These include:

1. Ignore the problem
2. Deny that the problem exists
3. Devise an alternative that circumvents the problem
4. Defer a solution to the problem
5. Confront the problem and find a solution to it

If the problem is to be confronted and solved, then a thorough analysis of the problem, from an economic analysis perspective, must be performed. The results of the analysis will indicate the specific approach that may be suitable in pursuing the innovation.

COST CONCEPTS IN INNOVATION

Cost management in a project environment refers to the functions required to maintain effective financial control of the project throughout its life cycle. There are several cost concepts that influence the economic aspects of managing projects. Within a given scope of analysis, there may be a combination of different types of cost aspects to consider. These cost aspects include the ones discussed in the following:

Actual Cost of Work Performed. This represents the cost actually incurred and recorded in accomplishing the work performed within a given time period.

Applied Direct Cost. This represents the amounts recognized in the time period associated with the consumption of labor, material, and other direct resources, without regard to the date of commitment or the date of payment. These amounts are to be charged to work in process (WIP) when resources are actually consumed, material resources are withdrawn from inventory for use, or material resources are received and scheduled for use within 60 days.

Budgeted Cost for Work Performed. This is the sum of the budgets for completed work plus the appropriate portion of the budgets for level of effort and apportioned effort. Apportioned effort is effort that by itself is not readily divisible into short-span work packages but is related in direct proportion to measured effort.

Budgeted Cost for Work Scheduled. This is the sum of budgets for all work packages and planning packages scheduled to be accomplished (including WIP) plus the amount of level of effort and apportioned effort scheduled to be accomplished within a given period of time.

Direct Cost. This is a cost that is directly associated with actual operations of a project. Typical sources of direct costs are direct material costs and direct labor costs. Direct costs are those that can be reasonably measured and allocated to a specific component of a project.

Economies of Scale. This refers to a reduction of the relative weight of the fixed cost in total cost by increasing output quantity. This helps to reduce the final unit cost of a product. Economies of scale is often simply referred to as the savings due to *mass production*.

Estimated Cost at Completion. This is the actual direct cost, plus indirect costs that can be allocated to the contract, plus the estimate of costs (direct and indirect) for authorized work remaining.

First Cost. This is the total initial investment required to initiate a project or the total initial cost of the equipment needed to start the project.

Fixed Cost. This is a cost incurred irrespective of the level of operation of a project. Fixed costs do not vary in proportion to the quantity of output. Example of costs that make up the fixed cost of a project are administrative expenses, certain types of taxes, insurance cost, depreciation cost, and debt servicing cost. These costs usually do not vary in proportion to quantity of output.

Incremental Cost. This refers to the additional cost of changing the production output from one level to another. Incremental costs are normally variable costs.

Indirect Cost. This is a cost that is indirectly associated with project operations. Indirect costs are those that are difficult to assign to specific components of a project. An example of an indirect cost is the cost of computer hardware and software needed to manage project operations. Indirect costs are usually calculated as a percentage of a component of direct costs. For example, the direct costs in an organization may be computed as 10% of direct labor costs.

Life Cycle Cost. This is the sum of all costs, recurring and nonrecurring, associated with a project during its entire life cycle.

Maintenance Cost. This is a cost that occurs intermittently or periodically for the purpose of keeping project equipment in good operating condition.

Marginal Cost. This is the additional cost of increasing production output by one additional unit. The marginal cost is equal to the slope of the total cost curve or line at the current operating level.

Operating Cost. This is a recurring cost needed to keep a project in operation during its life cycle. Operating costs may consist of such items as labor cost, material cost, and energy cost.

Opportunity Cost. This is the cost of forgoing the opportunity to invest in a venture that would have produced an economic advantage. Opportunity costs are usually incurred due to limited resources that make it impossible to take advantage of all investment opportunities. It is often defined as the cost of the best rejected opportunity. Opportunity costs can be incurred due to a missed opportunity rather than due to an intentional rejection. In many cases, opportunity costs are hidden or implied because they typically relate to future events that cannot be accurately predicted.

Overhead Cost. This is a cost incurred for activities performed in support of the operations of a project. The activities that generate overhead costs support the project efforts rather than contributing directly to the project goal. The handling of overhead costs varies widely from company to company. Typical overhead items are electric power cost, insurance premiums, cost of security, and inventory carrying cost.

Standard Cost. This is a cost that represents the normal or expected cost of a unit of the output of an operation. Standard costs are established in advance. They are developed as a composite of several component costs such as direct labor cost per unit, material cost per unit, and allowable overhead charge per unit.

Sunk Cost. This is a cost that occurred in the past and cannot be recovered under the present analysis. Sunk costs should have no bearing on the prevailing economic analysis and project decisions. Ignoring sunk costs is always a difficult task for analysts. For example, if $950,000 was spent four years ago to buy a piece of equipment for a technology-based project, a decision on whether or not to replace the equipment now should not consider that initial cost. But, uncompromising analysts might find it difficult to ignore that much money. Similarly, an individual making a decision on selling a personal automobile would typically try to relate the asking price to what was paid for the automobile when it was acquired. This is wrong under the strict concept of sunk costs.

Total Cost. This is the sum of all the variable and fixed costs associated with a project.

Variable Cost. This is a cost that varies in direct proportion to the level of operation or quantity of output. For example, the costs of material and labor required to make an item will be classified as variable costs because they vary with changes in level of output.

INNOVATION CASH FLOW ANALYSIS

The basic reason for performing economic analysis is to make a choice between mutually exclusive projects that are competing for limited resources. The cost performance of each project will depend on the timing and levels of its expenditures. The techniques of computing cash flow equivalence permit us to bring competing project cash flows to a common basis for comparison. The common basis depends on the prevailing interest rate. Two cash flows that are equivalent at a given interest rate will not be equivalent at a different interest rate. The basic techniques for converting cash flows from one point in time to another are presented in the next section.

TIME VALUE OF MONEY CALCULATIONS

Cash flow conversion involves the transfer of project funds from one point in time to another. The following notation is used for the variables involved in the conversion process:

i = interest rate per period
n = number of interest periods
P = a present sum of money
F = a future sum of money
A = a uniform end-of-period cash receipt or disbursement
G = a uniform arithmetic gradient increase in period-by-period payments or disbursements

In many cases, the interest rate used in performing economic analysis is set equal to the minimum attractive rate of return (MARR) of the decision-maker. The MARR is also sometimes referred to as *hurdle rate, required internal rate of return* (IRR), *return on investment* (ROI), or *discount rate*. The value of MARR is chosen with the objective of maximizing the economic performance of a project.

Compound Amount Factor. The procedure for the single payment compound amount factor finds a future sum of money, F, that is equivalent to a present sum of money, P, at a specified interest rate, i, after n periods. This is calculated as

$$F = P(1 + i)^n$$

Example. A sum of $5,000 is deposited in a project account and left there to earn interest for 15 years. If the interest rate per year is 12%, the compound amount after 15 years can be calculated as follows:

$$F = \$5,000(1 + 0.12)^{15} = \$27,367.85$$

Present Worth Factor. The present worth factor computes P when F is given. The present worth factor is obtained by solving for P in the equation for the compound amount factor. That is,

$$P = F(1 + i)^{-n}$$

Suppose it is estimated that $15,000 would be needed to complete the implementation of a project five years from now. How much should be deposited in a special project fund now so that the fund would accrue to the required $15,000 exactly five years from now? If the special project fund pays interest at 9.2% per year, the required deposit would be

$$P = \$15,000(1 + 0.092)^{-5} = \$9,660.03$$

Uniform Series Present Worth Factor. The uniform series present worth factor is used to calculate the present worth equivalent, P, of a series of equal end-of-period amounts, A. The derivation of the formula uses the finite sum of the present values of the individual amounts in the uniform series cash flow as shown in the following. Some formulas for series and summation operations are presented in the Appendix at the end of the book.

$$P = \sum_{t=1}^{n} A(1+i)^{-1} = A\left[\frac{(1+i)^n - 1}{i(1+i)^n}\right]$$

Example. Suppose the sum of $12,000 must be withdrawn from an account to meet the annual operating expenses of a multiyear project. The project account pays interest at 7.5% per year compounded on an annual basis. If the project is expected to last ten years, how much must be deposited in the project account now so that the operating expenses of $12,000 can be withdrawn at the end of every year for ten years? The project fund is expected to be depleted to zero by the end of the last year of the project. The first withdrawal will be made one year after the project account is opened, and no additional deposits will be made in the account during the project life cycle. The required deposit is calculated to be

$$P = \$12,000\left[\frac{(1+0.075)^{10} - 1}{0.075(1+0.075)^{10}}\right] = \$82,368.92$$

Uniform Series Capital Recovery Factor. The capital recovery formula is used to calculate the uniform series of equal end-of-period payments, A, that are equivalent to a given present amount, P. This is the converse of the uniform series present amount factor. The equation for the uniform series capital recovery factor is obtained by solving for A in the uniform series present amount factor. That is,

$$A = P\left[\frac{i(1+i)^n}{(1+i)^n - 1}\right]$$

Example. Suppose a piece of equipment needed to launch a project must be purchased at a cost of $50,000. The entire cost is to be financed at 13.5% per year and repaid on a monthly installment schedule over four years. It is desired to calculate what the monthly loan payments will be. It is assumed that the first loan payment will be made exactly one month after the equipment is financed. If the interest rate of 13.5% per year is compounded monthly, then the interest rate per month will be 13.5%/12 = 1.125% per month. The number of interest periods over which the loan will be repaid is 4(12) = 48 months. Consequently, the monthly loan payments are calculated to be

$$A = \$50,000 \left[\frac{0.01125(1+0.01125)^{48}}{(1+0.01125)^{48} - 1} \right] = \$1,353.82$$

Uniform Series Compound Amount Factor. The series compound amount factor is used to calculate a single future amount that is equivalent to a uniform series of equal end-of-period payments. Note that the future amount occurs at the same point in time as the last amount in the uniform series of payments. The factor is derived as shown here:

$$F = \sum_{t=1}^{n} A(1+i)^{n-1} = A \left[\frac{(1+i)^n - 1}{i} \right]$$

Example. If equal end-of-year deposits of $5,000 are made to a project fund paying 8% per year for ten years, how much can be expected to be available for withdrawal from the account for capital expenditure immediately after the last deposit is made?

$$F = \$5000 \left[\frac{(1+0.08)^{10} - 1}{0.08} \right] = \$72,432.50$$

Uniform Series Sinking Fund Factor. The sinking fund factor is used to calculate the uniform series of equal end-of-period amounts, A, that are equivalent to a single future amount, F. This is the reverse of the uniform series compound amount factor. The formula for the sinking fund is obtained by solving for A in the formula for the uniform series compound amount factor. That is,

$$A = F \left[\frac{i}{(1+i)^n - 1} \right]$$

Example. How large are the end-of-year equal amounts that must be deposited into a project account so that a balance of $75,000 will be available for withdrawal immediately after the 12th annual deposit is made? The initial balance in the account is zero at the beginning of the first year. The account pays 10% interest per year. Using the formula for the sinking fund factor, the required annual deposits are

$$A = \$75,000 \left[\frac{0.10}{(1+0.10)^{12} - 1} \right] = \$3,507.25$$

Capitalized Cost Formula. Capitalized cost refers to the present value of a single amount that is equivalent to a perpetual series of equal end-of-period payments. This is an extension of the series present worth factor with an infinitely large number of periods.

Using the limit theorem from calculus as n approaches infinity, the series present worth factor reduces to the following formula for the capitalized cost:

$$P = \lim_{n\to\infty} A\left[\frac{(1+i)^n - 1}{i(1+i)^n}\right] = A\left\{\lim_{n\to\infty}\left[\frac{(1+i)^n - 1}{i(1+i)^n}\right]\right\} = A\left(\frac{1}{i}\right)$$

Example. How much should be deposited in a general fund to service a recurring public service project to the tune of $6,500 per year forever if the fund yields an annual interest rate of 11%? Using the capitalized cost formula, the required one-time deposit to the general fund is

$$P = \frac{\$6,500}{0.11} = \$59,090.91$$

The formulas presented represent the basic cash flow conversion factors. The factors are widely tabulated, for convenience, in engineering economy books. Several variations and extensions of the factors are available. Such extensions include the arithmetic gradient series factor and the geometric series factor. Variations in the cash flow profiles include situations where payments are made at the beginning of each period rather than at the end and situations where a series of payments contains unequal amounts. Conversion formulas can be derived mathematically for those special cases by using the basic factors presented earlier.

Arithmetic Gradient Series. The gradient series cash flow involves an increase of a fixed amount in the cash flow at the end of each period. Thus, the amount at a given point in time is greater than the amount at the preceding period by a constant amount. This constant amount is denoted by G. The size of the cash flow in the gradient series at the end of period t is calculated as

$$A_t = (t-1)G, \quad t = 1,2,...,n$$

The total present value of the gradient series is calculated by using the present amount factor to convert each individual amount from time t to time 0 at an interest rate of $i\%$ per period and summing up the resulting present values. The finite summation reduces to a closed form as shown here:

$$P = \sum_{t=1}^{n} A_t(1+i)^{-t} = \sum_{t=1}^{n}(t-1)G(1+i)^{-t} = G\sum_{t=1}^{n}(t-1)(1+i)^{-t}$$

$$= G\left[\frac{(1+i)^n - (1+ni)}{i^2(1+i)^n}\right]$$

Example. The cost of supplies for a ten-year period increases by $1,500 every year starting at the end of year two. There is no supplies cost at the end of the first year. If the interest rate is 8% per year, determine the present amount that must be set aside at time zero to take care of all the future supplies expenditures. We have $G = 1,500$, $i = 0.08$, and $n = 10$. Using the arithmetic gradient formula, we obtain

$$P = 1,500 \left[\frac{1 - (1 + 10(0.08))(1 + 0.08)^{-10}}{(0.08)^2} \right] = \$1,500(25.9768)$$

$$= \$38,965.20$$

In many cases, an arithmetic gradient starts with some base amount at the end of the first period and then increases by a constant amount thereafter. The nonzero base amount is denoted as A_1.

The calculation of the present amount for such cash flows requires breaking the cash flow into a uniform series cash flow of amount A_1 and an arithmetic gradient cash flow with zero base amount. The uniform series present worth formula is used to calculate the present worth of the uniform series portion, while the basic gradient series formula is used to calculate the gradient portion. The overall present worth is then calculated as

$$P = P_{\text{uniform series}} + P_{\text{gradient series}} = A_1 \left[\frac{(1+i)^n - 1}{i(1+i)^n} \right] + G \left[\frac{(1+i)^n - (1+ni)}{i^2(1+i)^n} \right]$$

Increasing Geometric Series Cash Flow. In an increasing geometric series cash flow, the amounts in the cash flow increase by a constant percentage from period to period. There is a positive base amount, A_1, at the end of period 1. The amount at time t is denoted as

$$A_t = A_{t-1}(1 + j), \quad t = 2, 3, \ldots, n$$

where j is the percentage increase in the cash flow from period to period. By doing a series of back substitutions, we can represent A_t in terms of A_1 instead of in terms of A_{t-1}, as shown here:

$$A_2 = A_1(1 + j)$$

$$A_3 = A_2(1 + j) = A_1(1 + j)(1 + j)$$

$$\ldots$$

$$A_t = A_1(1 + j)^{t-1}, \quad t = 1, 2, 3, \ldots, n$$

The formula for calculating the present worth of the increasing geometric series cash flow is derived by summing the present values of the individual cash flow amounts. That is,

$$P = \sum_{t=1}^{n} A_t (1+i)^{-t}$$

$$= \sum_{t=1}^{n} \left[A_1 (1+j)^{t-1} \right] (1+i)^{-t}$$

$$= \frac{A_1}{(1+j)} \sum_{t=1}^{n} \left(\frac{1+j}{1+i} \right)^t$$

$$= A_1 \left[\frac{1-(1+j)^n (1+i)^{-n}}{i-j} \right], \quad i \neq j$$

If $i = j$, this formula reduces to the limit as $i \to j$, shown here:

$$P = \frac{nA_1}{1+i}, \quad i = j$$

Example. Suppose funding for a five-year project is to increase by 6% every year with an initial funding of \$20,000 at the end of the first year. Determine how much must be deposited into a budget account at time zero in order to cover the anticipated funding levels if the budget account pays 10% interest per year. We have $j = 6\%$, $i = 10\%$, $n = 5$, $A_1 = \$20,000$. Therefore,

$$P = 20,000 \left[\frac{1-(1+0.06)^5 (1+0.10)^{-5}}{0.10-0.06} \right] = \$20,000(4.2267) = \$84,533.60$$

Decreasing Geometric Series Cash Flow. In a decreasing geometric series cash flow, the amounts in the cash flow decrease by a constant percentage from period to period. The cash flow starts at some positive base amount, A_1, at the end of period 1. The amount of time t is denoted as

$$A_t = A_{t-1}(1-j), \quad t = 2,3,...,n$$

where j is the percentage decrease in the cash flow from period to period. As in the case of the increasing geometric series, we can represent A_t in terms of A_1:

$$A_2 = A_1(1-j)$$

$$A_3 = A_2(1-j) = A_1(1-j)(1-j)$$

...

$$A_t = A_1(1-j)^{t-1}, \quad t = 1,2,3,...,n$$

The formula for calculating the present worth of the decreasing geometric series cash flow is derived by finite summation as in the case of the increasing geometric series. The final formula is

$$P = A_1 \left[\frac{1-(1-j)^n(1+i)^{-n}}{i+j} \right]$$

Example. The contract amount for a three-year project is expected to decrease by 10% every year with an initial contract of $100,000 at the end of the first year. Determine how much must be available in a contract reservoir fund at time zero in order to cover the contract amounts. The fund pays 10% interest per year. Because $j = 10\%$, $i = 10\%$, $n = 3$, and $A_1 = \$100,000$, we should have

$$P = 100,000 \left[\frac{1-(1-0.10)^3(1+0.10)^{-3}}{0.10+0.10} \right] = \$100,000(2.2615)$$

$$= \$226,150$$

Internal Rate of Return. The IRR for a cash flow is defined as the interest rate that equates the future worth at time n or present worth at time 0 of the cash flow to zero. If we let $i*$ denote the IRR, then we have

$$FW_{t=n} = \sum_{t=0}^{n} (\pm A_t)(1+i*)^{n-t} = 0$$

$$PW_{t=0} = \sum_{t=0}^{n} (\pm A_t)(1+i*)^{-t} = 0$$

where "+" is used in the summation for positive cash flow amounts or receipts and "−" is used for negative cash flow amounts or disbursements. A_t denotes the cash flow amount at time t, which may be a receipt (+) or a disbursement (−). The value of $i*$ is referred to as *discounted cash flow rate of return, internal rate of return,* or *true rate of return.* This procedure essentially calculates the net future worth or the net present worth of the cash flow. That is,

Net future worth = Future worth of receipts − Future worth of disbursements

$$NFW = FW_{receipts} - FW_{disbursements}$$

Net present worth = Present worth of receipts − Present worth of disbursements

$$NPW = PW_{receipts} - PW_{disbursements}$$

Setting the NPW or NFW equal to zero and solving for the unknown variable i determines the IRR of the cash flow.

Benefit–Cost Ratio. The benefit–cost ratio of a cash flow is the ratio of the present worth of benefits to the present worth of costs. This is defined as

$$B/C = \frac{\displaystyle\sum_{t=0}^{n} B_t(1+i)^{-t}}{\displaystyle\sum_{t=0}^{n} C_t(1+i)^{-t}} = \frac{PW_{benefits}}{PW_{costs}}$$

where B_t is the benefit (receipt) at time t and C_t is the cost (disbursement) at time t. If the benefit–cost ratio is greater than 1, then the investment is acceptable. If the ratio is less than 1, the investment is not acceptable. A ratio of 1 indicates a break-even situation for the project.

Simple Payback Period. Payback period refers to the length of time it will take to recover an initial investment. The approach does not consider the impact of the time value of money. Consequently, it is not an accurate method of evaluating the worth of an investment. However, it is a simple technique that is used widely to perform a "quick-and-dirty" assessment of investment performance. Also, the technique considers only the initial cost. Other costs that may occur after time zero are not included in the calculation. The payback period is defined as the smallest value of n (n_{min}) that satisfies the following expression:

$$\sum_{t=1}^{n_{min}} R_t \geq C_0$$

where R_t is the revenue at time t and C_0 is the initial investment. The procedure calls for a simple addition of the revenues period by period until enough total has been accumulated to offset the initial investment.

Example. An organization is considering installing a new computer system that will generate significant savings in material and labor requirements for order processing. The system has an initial cost of $50,000. It is expected to save the organization $20,000 a year. The system has an anticipated useful life of five years with a salvage value of $5,000. Determine how long it would take for the system to pay for itself from the savings it is expected to generate. Because the annual savings are uniform, we can calculate the payback period by simply dividing the initial cost by the annual savings. That is,

$$n_{min} = \frac{\$50,000}{\$20,000} = 2.5 \text{ years}$$

Note that the salvage value of $5,000 is not included in the calculation because the amount is not realized until the end of the useful life of the asset (i.e., after five years). In some cases, it may be desired to consider the salvage value. In that case, the amount to be offset by the annual savings will be the net cost of the asset. In that case, we would have

$$n_{min} = \frac{\$50,000 - \$5,000}{\$20,000} = 2.25 \text{ years}$$

If there are tax liabilities associated with the annual savings, those liabilities must be deducted from the savings before calculating the payback period.

Discounted Payback Period. In this book, we introduce the *discounted payback period* approach, in which the revenues are reinvested at a certain interest rate. The payback period is determined when enough money has been accumulated at the given interest rate to offset the initial cost as well as other interim costs. In this case, the calculation is done by the following expression:

$$\sum_{t=1}^{n_{min}} R_t(1+i)^{n_{min}-1} \geq \sum_{t=0}^{n_{min}} C_t$$

Example. A new solar cell unit is to be installed in an office complex at an initial cost of $150,000. It is expected that the system will generate annual cost savings of $22,500 on the electricity bill. The solar cell unit will need to be overhauled every five years at a cost of $5,000 per overhaul. If the annual interest rate is 10%, find the *discounted payback period* for the solar cell unit considering the time value of money. The costs of overhaul are to be considered in calculating the discounted payback period.

Solution. Using the single payment compound amount factor for one period iteratively, the following solution is obtained.

Time period 1: $22,500
Time period 2: $22,500 + $22,500(1.10)1 = $47,250
Time period 3: $22,500 + $47,250(1.10)1 = $74,475
Time period 4: $22,500 + $74,475(1.10)1 = $104,422.50
Time period 5: $22,500 + $104,422.50(1.10)1 − $5,000 = $132,364.75
Time period 6: $22,500 + $132,364.75(1.10)1 = $168,101.23

The initial investment is $150,000. By the end of period 6, we have accumulated $168,101.23, more than the initial cost. Interpolating between period 5 and period 6, we obtain

$$n_{min} = 5 + \frac{150,000 - 132,364.75}{168,101.23 - 132,364.75}(6-5) = 5.49$$

That is, it will take 5.49 years, or 5 years and 6 months, to recover the initial investment.

Investment Life for Multiple Returns. The time it takes an amount to reach a certain multiple of its initial level is often of interest in many investment scenarios. The "Rule of 72" is one simple approach to calculating how long it will take an investment to double in value at a given interest rate per period. The Rule of 72 gives the following formula for estimating the doubling period:

$$n = \frac{72}{i}$$

where i is the interest rate expressed as a percentage. Referring to the single payment compound amount factor, we can set the future amount equal to twice the present amount and then solve for n, the number of periods. That is, $F = 2P$. Thus,

$$2P = P(1 + i)^n$$

Solving for n in this equation yields an expression for calculating the exact number of periods required to double P:

$$n = \frac{\ln(2)}{\ln(1+i)}$$

where i is the interest rate expressed in decimals. In the general case, for exact computation, the length of time it would take to accumulate m multiple of P is expressed as

$$n = \frac{\ln(m)}{\ln(1+i)}$$

where m is the desired multiple. For example, at an interest rate of 5% per year, the time it would take an amount, P, to double in value ($m = 2$) is 14.21 years. This, of course, assumes that the interest rate will remain constant throughout the planning horizon.

EFFECTS OF INFLATION

Inflation is a major player in financial and economic analyses of projects. Multiyear projects are particularly subject to the effects of inflation. Inflation can be defined as the decline in purchasing power of money.

Some of the most common causes of inflation are:

- Increase in amount of currency in circulation
- Shortage of consumer goods
- Escalation of the cost of production
- Arbitrary increase of prices by resellers

The general effects of inflation are felt in terms of increase in the prices of goods and decrease in the worth of currency. In cash flow analysis, ROI for a project will be affected by time value of money as well as inflation. The *real interest rate* (d) is defined as the desired rate of return in the absence of inflation. When we talk of "today's dollars" or "constant dollars," we are referring to the use of real interest rate. *Combined interest rate* (i) is the rate of return combining real interest rate and inflation rate. If we denote the *inflation rate* as j, then the relationship between the different rates can be expressed as

$$1 + i = (1 + d)(1 + j)$$

Thus, the combined interest rate can be expressed as

$$i = d + j + dj$$

Note that if $j = 0$ (i.e., no inflation), then $i = d$. We can also define *commodity escalation rate* (g) as the rate at which individual commodity prices escalate. This may be greater than or less than the overall inflation rate. In practice, several measures are used to convey inflationary effects. Some of these are *consumer price index*, *producer price index*, and *wholesale price index*. A *"market basket" rate* is defined as the estimate of inflation based on a weighted average of the annual rates of change in the costs of a wide range of representative commodities. A "then-current" cash flow is a cash flow that explicitly incorporates the impact of inflation. A "constant worth" cash flow is a cash flow that does not incorporate the effect of inflation. The real interest rate, d, is used for analyzing constant worth cash flows.

The then-current cash flow is the equivalent cash flow considering the effect of inflation. C_k is what it would take to buy a certain "basket" of goods after k time periods if there were no inflation. T_k is what it would take to buy the same "basket" in k time period if inflation is taken into account. For the constant worth cash flow, we have

$$C_k = T_0, k = 1, 2, \ldots, n$$

and for the then-current cash flow, we have

$$T_k = T_0(1 + j)^k, k = 1, 2, \ldots, n$$

where j is the inflation rate. If $C_k = T_0 = \$100$ under the constant worth cash flow, then we mean \$100 worth of buying power. If we are using the commodity escalation rate, g, then we will have

$$T_k = T_0(1 + g)^k, k = 1, 2, \ldots, n$$

Thus, a then-current cash flow may increase based on both a regular inflation rate (j) and a commodity escalation rate (g). We can convert a then-current cash flow to a constant worth cash flow by using the following relationship:

$$C_k = T_k(1 + j)^{-k}, k = 1, 2, \ldots, n$$

If we substitute T_k from the commodity escalation cash flow into this expression for C_k, we get

$$C_k = T_k(1+j)^{-k} = T_0(1+g)^k(1+j)^{-k}$$

$$= T_0[(1+g)/(1+j)]^k, k = 1, 2, \ldots, n$$

Note that if $g = 0$ and $j = 0$, then $C_k = T_0$. That is, there is no inflationary effect. We now define effective commodity escalation rate (v) as

$$v = [(1+g)/(1+j)] - 1$$

and we can express the commodity escalation rate (g) as

$$g = v + j + vj$$

Inflation can have a significant impact on the financial and economic aspects of a project. Inflation may be defined, in economic terms, as the increase in the amount of currency in circulation, resulting in a relatively high and sudden fall in its value. To a producer, inflation means a sudden increase in the cost of items that serve as inputs for the production process (equipment, labor, materials, etc.). To the retailer, inflation implies an imposed higher cost of finished products. To an ordinary citizen, inflation portends an unbearable escalation of prices of consumer goods. All these views are interrelated in a project management environment.

The amount of money supply, as a measure of a country's wealth, is controlled by the government. With no other choice, governments often feel impelled to create more money or credit to take care of old debts and pay for social programs. When money is generated at a faster rate than the growth of goods and services, it becomes a surplus commodity, and its value (purchasing power) will fall. This means that there will be too much money available to buy only a few goods and services. When the purchasing power of a currency falls, each individual in a product's life cycle has to dispense more of the currency in order to obtain the product. Some of the classic concepts of inflation are discussed here:

1. Increases in producer's costs are passed on to consumers. At each stage of the product's journey from producer to consumer, prices are escalated disproportionately in order to make a good profit. The overall increase in the product's price is directly proportional to the number of intermediaries it encounters on its way to the consumer. This type of inflation is called *cost-driven (or cost-push) inflation.*
2. Excessive spending power of consumers forces an upward trend in prices. This high spending power is usually achieved at the expense of savings. The law of supply and demand dictates that the more the demand, the higher the price. This type of inflation is known as *demand-driven (or demand-pull) inflation.*

3. Impact of international economic forces can induce inflation in a local economy. Trade imbalances and fluctuations in currency values are notable examples of international inflationary factors.

4. Increasing base wages of workers generate more disposable income and hence, higher demands for goods and services. The high demand, consequently, creates a pull on prices. Coupled with this, employers pass on the additional wage cost to consumers through higher prices. This type of inflation is, perhaps, the most difficult to solve because wages set by union contracts and prices set by producers almost never fall – at least, not permanently. This type of inflation may be referred to as *wage-driven (or wage-push) inflation.*

5. Easy availability of credit leads consumers to "buy now and pay later" and thereby, creates another loophole for inflation. This is a dangerous type of inflation because the credit not only pushes prices up but also leaves consumers with less money later on to pay for the credit. Eventually, many credits become uncollectible debts, which may then drive the economy into recession.

6. Deficit spending results in an increase in money supply and thereby, creates less room for each dollar to get around. The popular saying "a dollar does not go far anymore" simply refers to inflation in layman's terms. The different levels of inflation may be categorized as discussed in the following.

Mild Inflation. When inflation is mild (2–4%), the economy actually prospers. Producers strive to produce at full capacity in order to take advantage of the high prices to the consumer. Private investments tend to be brisk, and more jobs become available. However, the good fortune may only be temporary. Prompted by the prevailing success, employers are tempted to seek larger profits, and workers begin to ask for higher wages. They cite their employer's prosperous business as a reason to bargain for bigger shares of the business profit. Thus, we end up with a vicious cycle where the producer asks for higher prices, the unions ask for higher wages, and inflation starts an upward trend.

Moderate Inflation. Moderate inflation occurs when prices increase at 5–9%. Consumers start purchasing more as an edge against inflation. They would rather spend their money now than watch it decline further in purchasing power. The increased market activity serves to fuel further inflation.

Severe Inflation. Severe inflation is indicated by price escalations of 10% or more. Double-digit inflation implies that prices rise much faster than wages do. Debtors tend to be the ones who benefit from this level of inflation, because they repay debts with money that is less valuable then the money borrowed.

Hyperinflation. When each price increase signals an increase in wages and costs, which again sends prices further up, the economy has reached a stage of malignant, galloping inflation or hyperinflation. Rapid and uncontrollable inflation destroys the economy. The currency becomes economically useless, as the government prints it excessively to pay for obligations.

Inflation can affect any project in terms of raw materials procurement, salaries and wages, and/or cost tracking dilemma. Some effects are immediate and easily observable. Other effects are subtle and pervasive. Whatever form it takes, inflation must be taken into account in long-term project planning and control. Large projects may be adversely affected by the effects of inflation in terms of cost overruns and poor resource utilization. The level of inflation will determine the severity of the impact on projects.

BREAK-EVEN ANALYSIS

Break-even analysis refers to the determination of the balanced performance level where project income is equal to project expenditure. The total cost of an operation is expressed as the sum of the fixed and variable costs with respect to output quantity. That is,

$$TC(x) = FC + VC(x)$$

where x is the number of units produced, $TC(x)$ is the total cost of producing x units, FC is the total fixed cost, and $VC(x)$ is the total variable cost associated with producing x units. The total revenue resulting from the sale of x units is defined as

$$TR(x) = px$$

where p is the price per unit. The profit due to the production and sale of x units of the product is calculated as

$$P(x) = TR(x) - TC(x)$$

The break-even point of an operation is defined as the value of a given parameter that will result in neither profit nor loss. The parameter of interest may be the number of units produced, the number of hours of operation, the number of units of a resource type allocated, or any other measure of interest. At the break-even point, we have the following relationship:

$$TR(x) = TC(x) \text{ or } P(x) = 0$$

In some cases, there may be a known mathematical relationship between cost and the parameter of interest. For example, there may be a linear cost relationship between the total cost of a project and the number of units produced. The cost expressions facilitate straightforward break-even analysis. When two project alternatives are compared, the break-even point refers to the point of indifference between the two alternatives. The variable $x1$ represents the point where projects A and B are equally desirable, $x2$ represents where A and C are equally desirable, and $x3$ represents where B and C are equally desirable. The analysis shows that if we are operating below a production level of $x2$ units, then project C is the preferred project among the three. If we are operating at a level more than $x2$ units, then project A is the best choice.

Example. Three project alternatives are being considered for producing a new product. The required analysis involves determining which alternative should be selected on the basis of how many units of the product are produced per year. Based on past records, there is a known relationship between the number of units produced per year, x, and the net annual profit, $P(x)$, from each alternative. The level of production is expected to be between 0 and 250 units per year. The net annual profits (in thousands of dollars) are given here for each alternative:

$$\text{Project A: } P(x) = 3x - 200$$

$$\text{Project B: } P(x) = x$$

$$\text{Project C: } P(x) = (1/50)x^2 - 300$$

This problem can be solved mathematically by finding the intersection points of the profit functions and evaluating the respective profits over the given range of product units. It can also be solved by a graphical approach. Such a plot is called a *break-even chart*. The plot shows that Project B should be selected if between 0 and 100 units are to be produced. Project A should be selected if between 100 and 178.1 units (178 physical units) are to be produced. Project C should be selected if more than 178 units are to be produced. It should be noted that if fewer than 66.7 units (66 physical units) are produced, Project A will generate net loss rather than net profit. Similarly, Project C will generate losses if fewer than 122.5 units (122 physical units) are produced.

PROFIT RATIO ANALYSIS

Break-even charts offer opportunities for several different types of analysis. In addition to the break-even points, other measures of worth or criterion measures may be derived from the charts. A measure, called *profit ratio* is presented here for the purpose of obtaining a further comparative basis for competing projects. Profit ratio is defined as the ratio of the profit area to the sum of the profit and loss areas in a break-even chart. That is,

$$\text{Profit ratio} = \frac{\text{Area of profit region}}{\text{Area of profit region} + \text{Area of loss region}}$$

For example, suppose the expected revenue and the expected total cost associated with a project are given, respectively, by the following expressions:

$$R(x) = 100 + 10x$$

$$TC(x) = 2.5x + 250$$

where x is the number of units produced and sold from the project. The break-even point is shown to be 20 units. Net profits are realized from the project if more than 20 units are produced, and net losses are realized if fewer than 20 units are produced. It should be noted that the revenue function represents an unusual case where a revenue of $100 is realized when zero units are produced.

Suppose it is desired to calculate the profit ratio for this project if the number of units that can be produced is limited to between 0 and 100 units. The surface area of the profit region and the area of the loss region can be calculated by using the standard formula for finding the area of a triangle: Area = (1/2)(Base)(Height). Using this formula, we have the following:

$$\text{Area of profit region} = \frac{1}{2}\left(\text{Base}\right)\left(\text{Height}\right)$$

$$= \frac{1}{2}\left(1,100 - 500\right)\left(100 - 20\right)$$

$$= 24,000 \text{ square units}$$

$$\text{Area of loss region} = \frac{1}{2}\left(\text{Base}\right)\left(\text{Height}\right)$$

$$= \frac{1}{2}\left(250 - 100\right)\left(20\right)$$

$$= 1,500 \text{ square units}$$

Thus, the profit ratio is computed as

$$\text{Profit ratio} = \frac{24,000}{24,000 + 1,500}$$

$$= 0.9411$$

$$= 94.11\%$$

The profit ratio may be used as a criterion for selecting among project alternatives. If this is done, the profit ratios for all the alternatives must be calculated over the same values of the independent variable. The project with the highest profit ratio will be selected as the desired project. Both the revenue and cost functions for the project are nonlinear. The revenue and cost are defined as follows:

$$R(x) = 160x - x^2$$

$$TC(x) = 500 + x^2$$

If the cost and/or revenue functions for a project are not linear, the areas bounded by the functions may not be easily determined. For those cases, it may be necessary to use techniques such as definite integrals to find the areas. The computations indicate that the project generates a loss if fewer than 3.3 units (3 actual units) or more than

76.8 (76 actual units) are produced. The respective profit and loss areas on the chart are calculated as follows:

$$\text{Area 1 (loss)} \quad = \int_0^{3.3}\left[\left(500+x^2\right)-\left(160x-x^2\right)\right]dx$$

$$= 802.8 \text{ unit-dollars}$$

$$\text{Area 2 (profit)} = \int_{3.3}^{76.8}\left[\left(160x-x^2\right)-\left(500+x^2\right)\right]dx$$

$$= 132,272.08 \text{ unit-dollars}$$

$$\text{Area 3(loss)} \quad = \int_{76.8}^{100}\left[\left(500+x^2\right)-\left(160x-x^2\right)\right]dx$$

$$= 48,135.98 \text{ unit-dollars}$$

Consequently, the profit ratio for Project II is computed as

$$\text{Profit ratio} = \frac{\text{Total area of profit region}}{\text{Total area of profit region} + \text{Total area of loss region}}$$

$$= \frac{132,272.08}{802.76+132,272.08+48,135.98}$$

$$= 0.7299$$

$$= 72.99\%$$

The profit ratio approach evaluates the performance of each alternative over a specified range of operating levels. Most of the existing evaluation methods use single-point analysis with the assumption that the operating condition is fixed at a given production level. The profit ratio measure allows an analyst to evaluate the net yield of an alternative given that the production level may shift from one level to another. An alternative, for example, may operate at a loss for most of its early life, while it may generate large incomes to offset the losses in its later stages. Conventional methods cannot easily capture this type of transition from one performance level to another. In addition to being used to compare alternate projects, the profit ratio may also be used for evaluating the economic feasibility of a single project. In such a case, a decision rule may be developed. An example of such a decision rule is:

If profit ratio is greater than 75%, accept the project
If profit ratio is less than or equal to 75%, reject the project

AMORTIZATION ANALYSIS FOR INNOVATION INVESTMENT

Many capital investment projects are financed with external funds. A careful analysis must be conducted to ensure that the amortization schedule can be handled by the organization involved. A computer program such as GAMPS (graphic evaluation of amortization payments) might be used for this purpose. The program analyzes the installment payments, the unpaid balance, principal amounts paid per period, total installment payment, and current cumulative equity. It also calculates the "equity break-even point" (Badiru, 2016) for the debt being analyzed. The equity break-even point indicates the time when the unpaid balance on a loan is equal to the cumulative equity on the loan. With the output of this program, the basic cost of servicing the project debt can be evaluated quickly. A part of the output of the program presents the percentage of the installment payment going into equity and interest charge, respectively. The computational procedure for analyzing project debt follows these steps:

1. Given a principal amount, P, a periodic interest rate, i (in decimals), and a discrete time span of n periods, the uniform series of equal end-of-period payments needed to amortize P is computed as

$$A = \frac{P[i(1+i)^n]}{(1+i)^n - 1}$$

It is assumed that the loan is to be repaid in equal monthly payments. Thus, $A(t) = A$ for each period t throughout the life of the loan.

2. The unpaid balance after making t installment payments is given by

$$U(t) = \frac{A[1-(1+i)^{t-n}]}{i}$$

3. The amount of equity or principal amount paid with installment payment number t is given by

$$E(t) = A(1+i)^{t-n-1}$$

4. The amount of interest charge contained in installment payment number t is derived to be

$$I(t) = A[1-(1+i)^{t-n-1}]$$

where $A = E(t) + I(t)$.

5. The cumulative total payment made after t periods is denoted by

$$C(t) = \sum_{k=1}^{t} A(k) = \sum_{k=1}^{t} A = (A)(t)$$

6. The cumulative interest payment after t periods is given by

$$Q(t) = \sum_{x=1}^{t} I(x)$$

7. The cumulative principal payment after t periods is computed as

$$S(t) = \sum_{k=1}^{t} E(k) = A \sum_{k=1}^{t} (1+i)^{-(n-k+1)} = A\left[\frac{(1+i)^t - 1}{i(1+i)^n}\right]$$

where

$$\sum_{n-1}^{t} x^n = \frac{x^{t+1} - x}{x - 1}$$

8. The percentage of interest charge contained in installment payment number t is

$$f(t) = \frac{I(t)}{A}(100\%)$$

9. The percentage of cumulative interest charge contained in the cumulative total payment up to and including payment number t is

$$F(t) = \frac{Q(t)}{C(t)}(100\%)$$

10. The percentage of cumulative principal payment contained in the cumulative total payment up to and including payment number t is

$$H(t) = \frac{S(t)}{C(t)} = \frac{C(t) - Q(t)}{C(t)} = 1 - \frac{Q(t)}{C(t)} = 1 - F(t)$$

Example. Suppose that a manufacturing productivity improvement project is to be financed by borrowing $500,000 from an industrial development bank. The annual nominal interest rate for the loan is 10%. The loan is to be repaid in equal monthly installments over a period of 15 years. The first payment on the loan is to be made exactly one month after financing is approved. It is desired to perform a detailed analysis of the loan schedule.

The tabulated result shows a monthly payment of $5,373.04 on the loan. Considering time $t = 10$ months, one can see the following results:

$U(10) = \$487,475.13$ (unpaid balance)

$A(10) = \$5,373.04$ (monthly payment)

$E(10) = \$1,299.91$ (equity portion of the tenth payment)

$I(10) = \$4,073.13$ (interest charge contained in the tenth payment)

$C(10) = \$53,730.40$ (total payment to date)

$S(10) = \$12,526.21$ (total equity to date)

$f(10) = 75.81\%$ (percentage of the tenth payment going into interest charge)

$F(10) = 76.69\%$ (percentage of the total payment going into interest charge)

Thus, over 76% of the sum of the first ten installment payments goes into interest charges. The analysis shows that by time $t = 180$, the unpaid balance has been reduced to zero. That is, $U(180) = 0.0$. The total payment made on the loan is $967,148.40, and the total interest charge is $967,148.20 − $500,000 = $467,148.20. Thus, 48.30% of the total payment goes into interest charges. The information about interest charges might be very useful for tax purposes. The tabulated output shows that equity builds up slowly, while the unpaid balance decreases slowly. Note that very little equity is accumulated during the first three years of the loan schedule. The effects of inflation, depreciation, property appreciation, and other economic factors are not included in the analysis presented here. A project analyst should include such factors whenever they are relevant to the loan situation.

The point at which the curves intersect is referred to as the *equity break-even point*. It indicates when the unpaid balance is exactly equal to the accumulated equity or the cumulative principal payment. For the example, the equity break-even point is 120.9 months (over 10 years). The importance of the equity break-even point is that any equity accumulated after that point represents the amount of ownership or equity that the debtor is entitled to after the unpaid balance on the loan is settled with project collateral. The implication of this is very important, particularly in the case of mortgage loans. "Mortgage" is a word with French origin, meaning *death pledge* – perhaps a sarcastic reference to the burden of mortgage loans. The equity break-even point can be calculated directly from the formula derived as follows:

Let the equity break-even point, x, be defined as the point where $U(x) = S(x)$. That is,

$$A\left[\frac{1-(1+i)^{-(n-x)}}{i}\right] = A\left[\frac{(1+i)^x - 1}{i(1+i)^n}\right]$$

Multiplying both the numerator and denominator of the left-hand side of this expression by $(1 + i)^n$ and simplifying yields

$$\frac{(1+i)^n - (1+i)^x}{i(1+i)^n}$$

on the left-hand side. Consequently, we have

$$(1+i)^n - (1+i)^x = (1+i)^x - 1$$

$$(1+i)^x = \frac{(1+i)^n + 1}{2}$$

which yields the equity break-even expression:

$$x = \frac{\ln[0.5(1+i)^n + 0.5]}{\ln(1+i)}$$

where

ln is the natural log function
n is the number of periods in the life of the loan
i is the interest rate per period

The total payment starts from $0.0 at time 0 and goes up to $967,147.20 by the end of the last month of the installment payments. Because only $500,000 was borrowed, the total interest payment on the loan is $967,147.20 − $500,000 = $467,147.20. The cumulative principal payment starts at $0.0 at time 0 and slowly builds up to $500,001.34, which is the original loan amount. The extra $1.34 is due to round-off error in the calculations.

The percentage of interest charge in the monthly payments starts at 77.55% for the first month and decreases to 0.83% for the last month. By comparison, the percentage of interest in the total payment also starts at 77.55% for the first month and slowly decreases to 48.30% by the time the last payment is made at time 180. It is noted that an increasing proportion of the monthly payment goes into the principal payment as time goes on. If the interest charges are tax deductible, the decreasing values of $f(t)$ mean that there would be decreasing tax benefits from the interest charges in the later months of the loan.

INNOVATION COST ESTIMATION

Cost estimation and budgeting help establish a strategy for allocating resources in project planning and control. There are three major categories of cost estimation for budgeting. These are based on the desired level of accuracy. The categories are *order-of-magnitude estimates*, *preliminary cost estimates*, and *detailed cost estimates*. Order-of-magnitude cost estimates are usually gross estimates based on the experience and judgment of the estimator. They are sometimes called "ballpark" figures. These estimates are typically made without a formal evaluation of the details involved in the project. The level of accuracy associated with order-of-magnitude estimates can range from −50% to +50% of the actual cost. These estimates provide a quick way of getting cost information during the initial stages of a project.

50%(Actual cost) ≤ Order-of-magnitude estimate ≤ 150%(Actual cost)

Preliminary cost estimates are also gross estimates, but with a higher level of accuracy. In developing preliminary cost estimates, more attention is paid to some selected details of the project. An example of a preliminary cost estimate is the estimation of expected labor cost. Preliminary estimates are useful for evaluating project alternatives before final commitments are made. The level of accuracy associated with preliminary estimates can range from −20% to +20% of the actual cost.

80%(Actual cost) ≤ Preliminary estimate ≤ 120%(Actual cost)

Detailed cost estimates are developed after careful consideration has been given to all the major details of a project. Considerable time is typically needed to obtain detailed cost estimates. Because of the amount of time and effort needed to develop detailed cost estimates, the estimates are usually developed after there is firm commitment that the project will take off. Detailed cost estimates are important for evaluating actual cost performance during the project. The level of accuracy associated with detailed estimates normally ranges from −5% to +5% of the actual cost.

95%(Actual cost) ≤ Detailed cost ≤ 105%(Actual cost)

There are two basic approaches to generating cost estimates. The first one is a variant approach, in which cost estimates are based on variations of previous cost records. The other approach is the generative cost estimation, in which cost estimates are developed from scratch without taking previous cost records into consideration.

Optimistic and Pessimistic Cost Estimates. Using an adaptation of the program evaluation and review technique (PERT) formula, we can combine optimistic and pessimistic cost estimates. Let:

O = optimistic cost estimate
M = most likely cost estimate
P = pessimistic cost estimate

Then, the estimated cost can be estimated as

$$E[C] = \frac{O + 4M + P}{6}$$

and the cost variance can be estimated as

$$V[C] = \left[\frac{P - O}{6} \right]^2$$

BUDGETING AND CAPITAL ALLOCATION

Budgeting involves sharing limited resources between several project groups or functions in a project environment. Budget analysis can serve any of the following purposes:

- A plan for resources expenditure
- A project selection criterion
- A projection of project policy
- A basis for project control
- A performance measure
- A standardization of resource allocation
- An incentive for improvement

Top-Down Budgeting. Top-down budgeting involves collecting data from upper-level sources such as top and middle managers. The numbers supplied by the managers may come from their personal judgment, past experience, or past data on similar project activities. The cost estimates are passed to lower-level managers, who then break the estimates down into specific work components within the project. These estimates may, in turn, be given to line managers, supervisors, and lead workers to continue the process until individual activity costs are obtained. Top management provides the global budget, while the functional-level worker provides specific budget requirements for the project items.

Bottom-up Budgeting. In this method, elemental activities and their schedules, descriptions, and labor skill requirements are used to construct detailed budget requests. Line workers familiar with specific activities are requested to provide cost estimates. Estimates are made for each activity in terms of labor time, materials, and machine time. The estimates are then converted to an appropriate cost basis. The dollar estimates are combined into composite budgets at each successive level up the budgeting hierarchy. If estimate discrepancies develop, they can be resolved through the intervention of senior management, middle management, functional managers, project manager, accountants, or standard cost consultants.

Elemental budgets may be developed on the basis of time-based incremental progress of the project. When all the individual estimates are gathered, we obtain a composite budget estimate. The bar chart appended to a segment of the pie chart indicates the individual cost components making up that particular segment. Analytical tools such as learning curve analysis, work sampling, and statistical estimation may be employed in the cost estimation and budgeting processes.

Mathematical Formulation of Capital Allocation. Capital rationing involves selecting a combination of projects that will optimize the ROI. A mathematical formulation of the capital budgeting problem is presented in the follows:

$$\text{Maximize } z = \sum_{i=1}^{n} v_i x_i$$

$$\text{Subject to } \sum_{i=1}^{n} c_i x_i \leq B$$

$x_i = 0, 1; i = 1, \ldots, n$
   where

   n = number of projects
   v_i = measure of performance for project i (e.g., present value)
   c_i = cost of project i
   x_i = indicator variable for project i
   B = budget availability level

A solution of this model will indicate which projects should be selected in combination with which projects. The example that follows illustrates a capital rationing problem.

Example of a Capital Rationing Problem. Planning of a portfolio of projects is essential in resource-limited projects. The capital rationing example presented here involves the determination of the optimal combination of project investments so as to maximize total ROI. Suppose that a project analyst is given N projects, $X_1, X_2, X_3, \ldots$, X_N, with the requirement to determine the level of investment in each project so that total investment return is maximized subject to a specified limit on available budget. The projects are not mutually exclusive.

The investment in each project starts at a base level b_i ($i = 1, 2, \ldots, N$) and increases by a variable increment k_{ij} ($j = 1, 2, 3, \ldots, K_i$), where K_i is the number of increments used for project i. Consequently, the level of investment in project X_i is defined as

$$x_i = b_i + \sum_{j=1}^{K_i} k_{ij}$$

where

$$x_i \geq 0 \qquad \forall i$$

For most cases, the base investment will be zero. In those cases, we will have $b_i = 0$. In the modeling procedure used for this problem, we have

$$X_i = \begin{cases} 1 & \text{if the investment in project } i \text{ is greater than zero} \\ 0 & \text{otherwise} \end{cases}$$

and

$$Y_{ij} = \begin{cases} 1 & \text{if } j\text{th increment of alternative } i \text{ is used} \\ 0 & \text{otherwise} \end{cases}$$

The variable x_i is the actual level of investment in project i, while X_i is an indicator variable indicating whether or not project i is one of the projects selected for investment. Similarly, k_{ij} is the actual magnitude of the jth increment, while Y_{ij} is an indicator variable that indicates whether or not the jth increment is used for project i. The maximum possible investment in each project is defined as M_i, such that

$$b_i \leq x_i \leq M_i$$

There is a specified limit, B, on the total budget available to invest, such that

$$\sum_i x_i \leq B$$

There is a known relationship between the level of investment, x_i, in each project and the expected return, $R(x_i)$. This relationship will be referred to as the *utility function*, $f(\bullet)$, for the project. The utility function may be developed through historical data, regression analysis, and forecasting models. For a given project, the utility function is used to determine the expected return, $R(x_i)$, for a specified level of investment in that project. That is,

$$R(x_i) = f(x_i) = \sum_{j=1}^{K_i} r_{ij} Y_{ij}$$

where r_{ij} is the incremental return obtained when the investment in project i is increased by k_{ij}. If the incremental return decreases as the level of investment increases, the utility function will be *concave*. In that case, we will have the following relationship:

$$r_{ij} \geq r_{ij+1} \text{ or } r_{if} - r_{ij+1} \geq 0$$

Thus,

$$Y_{ij} \geq Y_{ij+1} \quad \text{or} \quad Y_{ij} - Y_{ij+1} \geq 0$$

so that only the first n increments ($j = 1, 2, \ldots, n$) that produce the highest returns are used for project i.

If the incremental returns do not define a concave function, $f(x_i)$, then one has to introduce the inequality constraints presented earlier into the optimization model. Otherwise, the inequality constraints may be left out of the model, since the first inequality, $Y_{ij} \geq Y_{ij+1}$, is always implicitly satisfied for concave functions. Our objective is to maximize the total return. That is,

$$\text{Maximize} \quad Z = \sum_i \sum_j r_{ij} Y_{ij}$$

Subject to the following constraints:

$$x_i = b_i + \sum_j k_{ij} Y_{ij} \quad \forall i$$

$$b_i \leq x_i \leq M_i \quad \forall i$$

$$Y_{ij} \geq Y_{ij+1} \quad \forall i, j$$

$$\sum_i x_i \leq B$$

$$x_i \geq 0 \quad \forall i$$

$$Y_{ij} = 0 \text{ or } 1 \quad \forall i, j$$

Now, suppose we are given four projects (i.e., $N = 4$) and a budget limit of $10 million.

For example, if an incremental investment of $0.20 million from stage 2 to stage 3 is made in project 1, the expected incremental return from the project will be $0.30 million. Thus, a total investment of $1.20 million in project 1 will yield a total return of $1.90 million.

The question addressed by the optimization model is to determine how many investment increments should be used for each project. That is, when should we stop increasing the investments in a given project? Obviously, for a single project, we would continue to invest as long as the incremental returns are larger than the incremental investments. However, for multiple projects, investment interactions complicate the decision, so that investment in one project cannot be independent of the other projects. The linear programming (LP) model of the capital rationing example was solved with LINDO software. The solution indicates the following values for Y_{ij}.

Project 1:

$$Y11 = 1, Y12 = 1, Y13 = 1, Y14 = 0, Y15 = 0$$

Thus, the investment in project 1 is X1 = $1.20 million. The corresponding return is $1.90 million.

Project 2:

$$Y21 = 1, Y22 = 1, Y23 = 1, Y24 = 1, Y25 = 0, Y26 = 0, Y27 = 0$$

Thus, the investment in project 2 is $X2 = \$3.80$ million. The corresponding return is $6.80 million.

Project 3:

$$Y31 = 1, Y32 = 1, Y33 = 1, Y34 = 1, Y35 = 0, Y36 = 0, Y37 = 0$$

Thus, the investment in project 3 is $X3 = \$2.60$ million. The corresponding return is $5.90 million.

Project 4:

$$Y41 = 1, Y42 = 1, Y43 = 1$$

Thus, the investment in project 4 is $X4 = \$2.35$ million. The corresponding return is $3.70 million.

The total investment in all four projects is $9,950,000. Thus, the optimal solution indicates that not all of the $10,000,000 available should be invested. The expected return from the total investment is $18,300,000. This translates into 83.92% ROI.

The optimal solution indicates an unusually large return on total investment. In a practical setting, expectations may need to be scaled down to fit the realities of the project environment. Not all optimization results will be directly applicable to real situations. Possible extensions of this model of capital rationing include the incorporations of risk and time value of money into the solution procedure. Risk analysis would be relevant, particularly for cases where the levels of returns for the various levels of investment are not known with certainty. The incorporation of time value of money would be useful if the investment analysis is to be performed for a given planning horizon. For example, we might need to make investment decisions to cover the next five years rather than just the current time.

INNOVATION COST MONITORING

As a project progresses, costs can be monitored and evaluated to identify areas of unacceptable cost performance. A plot of cost versus time for projected cost and actual cost can reveal a quick identification of when cost overruns occur in a project.

Plots similar to those presented earlier may be used to evaluate the cost, schedule, and time performance of a project. An approach similar to the profit ratio presented earlier may be used together with the plot to evaluate the overall cost performance of a project over a specified planning horizon. Presented below is a formula for *cost performance index (CPI)*:

$$CPI = \frac{\text{Area of cost benefit}}{\text{Area of cost benefit} + \text{Area of cost overrun}}$$

As in the case of the profit ratio, CPI may be used to evaluate the relative performance of several project alternatives or to evaluate the feasibility and acceptability of an individual alternative.

PROJECT BALANCE TECHNIQUE

One other approach to monitoring cost performance is the project balance technique. The technique helps in assessing the economic state of a project at a desired point in time in the life cycle of the project. It calculates the net cash flow of a project up to a given point in time. The project balance is calculated as

$$B(i)_t = S_t - P(1+i)^t + \sum_{k=1}^{t} PW_{income}(i)_k$$

where

$B(i)_i$ = project balance at time t at an interest rate of $i\%$ per period

$PW_{income}(i)_t$ = present worth of net income from the project up to time t

P = initial cost of the project

S_t = salvage value at time t

The project balance at time t gives the net loss or net project associated with the project up to that time.

COST CONTROL SYSTEM

Contract management involves the process by which goods and services are acquired, utilized, monitored, and controlled in a project. Contract management addresses the contractual relationships from the initiation of a project to the completion of the project (i.e., completion of services and/or handover of deliverables). Some of the important aspects of contract management are:

- Principles of contract law
- Bidding process and evaluation
- Contract and procurement strategies
- Selection of source and contractors
- Negotiation
- Worker safety considerations
- Product liability
- Uncertainty and risk management
- Conflict resolution

The following equations can be used to calculate cost and schedule variances for work packages at any point in time:

$$\text{Cost variance} = \text{BCWP} - \text{ACWP}$$

$$\text{Percent cost variance} = (\text{Cost variance}/\text{BCWP}) \cdot 100$$

$$\text{Schedule variance} = \text{BCWP} - \text{BCWS}$$

$$\text{Percent schedule variance} = (\text{Schedule variance}/\text{BCWS}) \cdot 100$$

$$\text{ACWP and remaining funds} = \text{Target cost (TC)}$$

$$\text{ACWP} + \text{cost to complete} = \text{Estimated cost at completion (EAC)}$$

SOURCES OF CAPITAL FOR INNOVATION

Financing a project means raising capital for the project. Capital is a resource consisting of funds available to execute a project. Capital includes not only privately owned production facilities but also public investment. Public investments provide the infrastructure of the economy, such as roads, bridges, water supply, and so on. Other public capital that indirectly supports production and private enterprise includes schools, police stations, central financial institutions, and postal facilities.

If the physical infrastructure of the economy is lacking, the incentive for private entrepreneurs to invest in production facilities is likely to be lacking also. The government or community leaders can create the atmosphere for free enterprise by constructing better roads, providing better public safety and facilities, and encouraging ventures that ensure adequate support services.

As far as project investment is concerned, what can be achieved with project capital is very important. The avenues for raising capital funds include banks, government loans or grants, business partners, cash reserves, and other financial institutions. The key to the success of the free enterprise system is the availability of capital funds and the availability of sources to invest the funds in ventures that yield products needed by the society. Some specific ways that funds can be made available for business investments are discussed here:

Commercial Loans. Commercial loans are the most common sources of project capital. Banks should be encouraged to loan money to entrepreneurs, particularly those just starting business. Government guarantees may be provided to make it easier for the enterprise to obtain the needed funds.

Bonds and Stocks. Bonds and stocks are also common sources of capital. National policies regarding the issuance of bonds and stocks can be developed to target specific project types to encourage entrepreneurs.

Interpersonal Loans. Interpersonal loans are unofficial means of raising capital. In some cases, there may be individuals with enough personal funds to provide personal loans to aspiring entrepreneurs. But presently, there is no official mechanism that handles the supervision of interpersonal business loans. If a supervisory body exists at a national level, wealthy citizens will be less apprehensive about loaning money to friends and relatives for business purposes. Thus, the wealthy citizens can become a strong source of business capital.

Foreign Investment. Foreign investments can be attracted for local enterprises through government incentives. The incentives may be in terms of attractive zoning permits, foreign exchange permits, or tax breaks.

Investment Banks. The operations of investment banks are often established to raise capital for specific projects. Investment banks buy securities from enterprises and resell them to other investors. Proceeds from these investments may serve as a source of business capital.

Mutual Funds. Mutual funds represent collective funds from a group of individuals. The collective funds are often large enough to provide capital for business investments. Mutual funds may be established by individuals or under the sponsorship of a government agency. Encouragement and support should be provided for the group to spend the money for business investment purposes.

Supporting Resources. A clearing house of potential goods and services that a new project can provide may be established by the government. New entrepreneurs interested in providing the goods and services should be encouraged to start relevant enterprises. They should be given access to technical, financial, and information resources to facilitate starting production operations.

The time value of money is an important factor in project planning and control. This is particularly crucial for long-term projects that are subject to changes in several cost parameters. Both the timing and quantity of cash flows are important for project management. The evaluation of a project alternative requires consideration of the initial investment, depreciation, taxes, inflation, economic life of the project, salvage value, and cash flows.

ACTIVITY-BASED COSTING

Activity-based costing (ABC) has emerged as an appealing costing technique in industry. The major motivation for ABC is that it offers an improved method to achieve enhancements in operational and strategic decisions. ABC offers a mechanism to allocate costs in direct proportion to the activities that are actually performed. This is an improvement over the traditional way of generically allocating costs to departments. It also improves the conventional approaches to allocating overhead costs.

The use of PERT/critical path method (CPM), precedence diagramming, and the recently developed approach of critical resource diagramming can facilitate task decomposition to provide information for ABC. Some of the potential impacts of ABC on a production line are:

- Identification and removal of unnecessary costs
- Identification of the cost impact of adding specific attributes to a product
- Indication of the incremental cost of improved quality
- Identification of the value-added points in a production process
- Inclusion of specific inventory carrying costs
- Provision of a basis for comparing production alternatives
- Ability to assess "what-if" scenarios for specific tasks

ABC is just one component of the overall activity-based management in an organization. Activity-based management involves a more global management approach to the planning and control of organizational endeavors. This requires consideration of product planning, resource allocation, productivity management, quality control, training, line balancing, value analysis, and a host of other organizational responsibilities. Thus, while ABC is important, one must not lose sight of the universality of the environment in which it is expected to operate. And frankly, there are some processes where functions are so intermingled that decomposition into specific activities may be difficult. Major considerations in the implementation of ABC are:

- Resources committed to developing activity-based information and cost
- Duration and level of effort needed to achieve ABC objectives
- Level of cost accuracy that can be achieved by ABC
- Ability to track activities based on ABC requirements
- Handling the volume of detailed information provided by ABC
- Sensitivity of the ABC system to changes in activity configuration

Income analysis can be enhanced by the ABC approach. Similarly, instead of allocating manufacturing overhead on the basis of direct labor costs, an ABC analysis could be performed. The specific ABC cost components can be further broken down if needed. A spreadsheet analysis would indicate the impact on net profit as specific cost elements are manipulated. Based on this analysis, it is seen that Product Line A is the most profitable. Product Line B comes in second even though it has the highest total line cost.

CONCLUSIONS

In any innovation enterprise, computations and analyses similar to those illustrated in this chapter are crucial in controlling and enhancing the bottom-line survival of the organization. Analysts can adapt and extend the techniques presented in this chapter for application to the prevailing scenarios in their respective organizations.

REFERENCES

Badiru, Adedeji B. (2016). Equity Break-even Point: A Graphical and Tabulation Tool for Engineering Managers. *Engineering Management Journal*, Vol. 28, No. 4, pp. 249–255.

Badiru, Adedeji B. (2019). *Project Management: Systems, Principles, and Applications*, 2nd edition, Taylor & Francis CRC Press, Boca Raton, FL.

Badiru, Adedeji B. and O. A. Omitaomu (2007). *Computational Economic Analysis for Engineering and Industry*, Taylor & Francis CRC Press, Boca Raton, FL.

Newnan, D. G., Ted G. Eschenbach, and Jerome P. Lavelle (2004). *Engineering Economic Analysis*, 9th edition, Oxford University Press, New York.

Sullivan, William G., Elin M. Wicks, and James T. Luxhoj (2003). *Engineering Economy*, 12th edition, Pearson Education, Inc., Upper Saddle River, NJ.

White, John A., Kellie S. Grasman, Kenneth E. Case, Kim LaScola Needy, and David B. Pratt (2014). *Fundamentals of Engineering Economic Analysis*, John Wiley, Hoboken, NJ.

6 Economic Metrics for Innovation

INTRODUCTION

Economic metrics are the most readily seen and understood in any innovation assessment. For this reason, an economic assessment represents the bottom line in innovation management with respect to several parameters of interest. This chapter examines the compatibility of systemic resilience, organizational robustness, and operational agility with effectiveness and efficiency as economic goals of defense organizations, endeavors, and programs. Its conceptual framework includes a discerning view of conventional economic instruments of performance measurement from the perspective of core competencies, critical infrastructure, supply security, and strategic relevance in a disruptive environment of crises, shocks, uncertainties, and adversities. The methodology involves benefit–cost analyses in light of viability evaluation based on quantitative as well as qualitative criteria. The research approach is designed to sensitize decision-makers dealing with conventional economic tools to comprehensively enmesh resilience metrics, particularly in an era of hybrid threats to national security. Thereby, the chapter examines viable instruments that support a strategic view on defense initiatives, activities, programs, endeavors, and projects to adequately address resilience aspects.

Unpredictability is one of the most appropriate designations of the systemic prerequisites our defense and security efforts primarily and constantly have to tackle. It proves fallacious to focus on merely one threat and risk dimension that modern armed forces are prepared to engage with. Climate change, conventional and non-conventional wars, riots and conflicts on all continents, weak and failing states, terrorism, proliferation of weapons of mass destruction, natural disasters, or pandemics illustrate the incredibly wide spectrum of potential global challenges and menaces to the striving for freedom, safety, wellbeing, and prosperity. Hybrid tactics and warfare scenarios are typically referred to as a combination of the engagement of traditional military operations with other coordinated destabilizing, intimidating, and misleading courses of action like disinformation campaigns, airspace violations, economic pressure, sabotage acts on the critical infrastructure, or cyberattacks. Modern societies offer various attack surfaces and persistently remain vulnerable to these pervasive threat potentials. The circumstance of unpredictable disruptiveness is to be taken into account in the advancement of defense capabilities.

With the modern understanding of preventive problem-solving, today's security policies focus on a comprehensive approach including diplomatic, military,

DOI: 10.1201/9781003403548-6

economic, and financial resources, development aid, rule-of-law building, and governance by both state institutions and civil organization. This comprehensive approach handles uncertainty and addresses complex problem domains by decisive responses, joint commitments, and alliances. Individual states or institutions can hardly deal with complex global threats on their own, especially in **hybrid scenarios**. The multifaceted, diffuse, and destructive impacts do not know national boundaries. Their responsible perpetrators can hide, be anonymous, be unknown, or deny involvement. They are often not clearly identifiable as state or private attackers. The most perfidious aspect is that the disruption can be caused without necessarily crossing the threshold to a state of war. In a comprehensive approach, defense against these 360° threats and risks requires **robust**, **agile**, and **resilient** defense structures and processes that are tightly intertwined with other government and civilian entities. To meet the required defense capabilities, it is crucial not only to build viability on effectiveness and efficiency criteria but also obligatorily to enmesh the **resilience metrics** in the analysis phase of measures. This procedure enables the following implementation and operational phases to succeed. In this regard, for the Organisation for Economic Co-operation and Development (OECD), *"resilience is a topic of increasing interest on an international scale"* (Linkov et al., 2018).

RESEARCH BACKGROUND: HYBRID THREATS AND RESILIENCE

The year 2022 was marked by numerous devastating weather events and natural disasters related to the exacerbating phenomenon of human-induced climate change. From wildfires and droughts in South Europe and disastrous floods in Pakistan, Nigeria, Australia, and South Africa, to severe heat waves in the USA and India, to name just a marginal excerpt of miscellaneous climate extremes, no region of the world was truly spared from the impacts of global warming (WRI, 2022; World Economic Forum, 2022). Russia's invasion of Ukraine, as the worst conventional war the European continent has seen since World War II, is not solely claiming innocent casualties, the sovereignty, freedom, and even the raison d'être of a whole nation, but also causing massive displacement and incredible damage to the livelihoods of millions of people, and also drastically impeding global energy and food supply (WRI, 2022; NATO, 2022a). It is doubtful that there is any world region that has not been affected by the direct or indirect ramifications of this war, through either sharply soaring inflation, food shortages, supply bottlenecks, or many others. In addition to that, Russia's ongoing massive bombardments on Ukraine's energy infrastructure reflect the vulnerability of ordinary daily life's needs, including heating, electricity, clean water, sewage, and telecommunications, while at the same time also revealing the strategic relevance of critical infrastructure in a modern conflict scenario (New York Times, 2022). The disruption triggered by the Covid-19 pandemic both to societies and to economic structures has not come to an end, even though constraints to regain public health and safety, such as lockdowns, have recently been considerably loosening around the world. A substantial recovery from the rigors of the pandemic is more difficult due to the adverse impacts sparked off by the invasion of Ukraine. A prominent and early example of cyberwarfare in the context of **hybrid threats** is the 2007 scenario in Estonia. Vast cyberattacks, just below the acknowledged level of an armed attack, accompanied by a massive exacerbation of false news and violent upheavals, resulted in major disconnections of media, financial, and governmental online services and eventually caused confusion, disturbance, and instability. The attacks raised awareness of the need for substantial cyberdefense capabilities and made the world cognizant that hostile online assaults may be covertly state backed. Furthermore, there have been recent dismaying incidents of sabotage on a large scale. In September 2022, several massive detonations on Nord Stream pipelines in the Baltic Sea, which leaked natural gas and disrupted the surface by up to 0.6 miles, constituted another severe case of critical infrastructure impairment, allegedly by sabotage. A further recent incident that highlights the prevalence of cyberattacks on critical infrastructure occurred in May 2021, when a ransomware attack crippled the pipeline system of a major gas, jet fuel, and diesel supplying company on the US East Coast (New York Times, 2021).

The common feature of these threat scenarios is the pervasive menace to **critical infrastructure** as the nation's central neural system. A further commonality is an environment of unpredictable, multidimensional disruption with impacts on society, government, military and civilian institutions, economy, finance, commerce, logistics, the judicial system, etc. From this understanding, an overriding security policy goal can be derived: to reduce vulnerability and to enhance robustness toward the multiple-risk environment. Cohesive, preventive, and resolute answers to tackle hybrid threats therefore require resilient procedures, structures, and processes. Ross et al. (2019) offer a pertinent definition of **resilience**:

> The ability to prepare for and adapt to changing conditions and withstand and recover rapidly from disruption. Resilience includes the ability to withstand and recover from deliberate attacks, accidents, or naturally occurring threats or incidents.

In summary, resilience is defined by resistivity toward adversity or uncertainty, and the ability to recover quickly from disruption. NATO has an equivalent concept that postulates: "each NATO member country needs to be resilient to resist and recover from a major shock such as a natural disaster, failure of critical infrastructure, or a hybrid or armed attack" (NATO, 2022b), referring to Article 3 of the NATO Treaty. We will apply this definitional basis in our scholarly approach.

Robustness is assumed to have a broader context than resilience and is defined by "underlying system mechanisms to proactively and agilely respond to predictable and unpredictable crises" (Desouza and Xie, 2021). Cîrdei (2018) understands robustness as the "ability to continue operating even under the circumstances of a serious accident or incident."

Harraf et al. (2015) describe **agility** as a "measure of responsiveness" to an external stimulus. Agility includes effectively planned or unplanned responses, the attributes of flexibility and adaptability, as well as "an organization's sensing, anticipating, entrepreneurial alertness and proactivity" (Harraf et al. 2015). As revealed in the US Air Force Future Operating Concept, in the defense context, the term **operational agility** applies. Operational agility encompasses "the ability to rapidly generate—and shift among—multiple solutions for a given challenge", or in a nutshell, "to act quickly in response to changing context" (U.S. Air Force, 2015).

ANALYSIS: INTEGRATION OF RESILIENCE ASPECTS

RELEVANT SPHERES OF RESILIENCE BUILD-UP

Strategically resilient defense capabilities require resilience efforts from every possible angle. **A top-down approach** with characteristics such as robust funding, comprehensive legislation, involvement of different stakeholders in crisis management, coordination of all relevant political disciplines, or civil–military collaboration, is relevant to enable resilient overall conditions on a basic level. The UK National Security Strategy, for instance, aims to achieve National Security Tasks to ensure a secure and resilient UK through tight coordination between the Government Departments and by the cooperation of intelligence capabilities with military, diplomatic, and domestic security requirements, and economic prosperity (UK Government, 2010).

To build, strengthen, and maintain national resilience and reduce collective vulnerability, a **bottom-up view** is of even higher significance. The society, the enterprises, the local authorities, public administration, fire departments, every single military base and unit, every household, the justice courts, the health facilities, etc., are all critical resilience "enablers" of a nation that carry the ability to endure crises, threats, vulnerabilities, risks, and adversities. The bottom line is that it is their mindset, preparedness, and behavior that determine whether robust, agile, and resilient structures and processes are fully operable in case of an emergency. Merely with a successful implementation of robustness on this level, the strengthening of national defense resilience is achievable in the first place. Under this premise, we target precisely this aspect and advocate the use of relevant techniques and methods to frame and structure the subject.

Critical infrastructure is the backdrop of every resilience consideration (Cîrdei, 2018). Disruption in this environment is immediately a substantial threat to national security and has to be regarded with the highest priority, hand in hand with measures to enhance resilience, robustness, and agility. Critical infrastructure is defined as a set of subsystems, assets, and strategic industries, physical or virtual, that are absolutely vital to a nation, such that the destruction of such subsystems, industries, or assets would have a debilitating impact on national security, economic vitality, public health, safety, or a combination of those considerations. According to the US Cybersecurity & Infrastructure Security Agency (CISA, 2022), there are 16 critical infrastructure sectors (listed here) that are vital to security, public health, economic security, or safety:

1. Emergency services sector
2. Transportation systems sector
3. Chemical sector

4. Commercial facilities sector
5. Communications sector
6. Critical manufacturing sector
7. Water and wastewater systems sector
8. Dams sector
9. Information technology sector
10. Defense industrial base sector
11. Healthcare and public health sector
12. Food and agriculture sector
13. Energy sector
14. Government facilities sector
15. Financial services sector
16. Nuclear services sector

If we take this US example of physical and virtual assets with the highest criticality for national security, we can derive where the main focus of reducing vulnerability is located, namely, on the sectors whose impairment would cause the most devastating impact on the country's functional capacity. A smart defense strategy addresses these drivers as a whole and puts the emphasis on bottom-up efforts to mitigate threats and risk in the form of enhanced resilience.

EFFECTIVENESS AND EFFICIENCY AS CONVENTIONAL ECONOMIC PARAMETERS

Effectiveness means an approach of maximum outcome or value achievement. It is highly compatible with the purpose of getting defense missions accomplished. To fulfill tasks with the best possible result, e.g., a full capture of an enemy stronghold by an airborne operation or depleting the combat power of a cyberattack unit as much as possible, typically characterizes the military mode of operation. Simultaneously, it is a common goal formulation in government programs. For instance, the Department of Defense (DoD) Directive 2311.01 *"Law of War Program"* states the implementation of *effective* programs to prevent violations of the law of war within the chain of command (DoD, 2020). Badiru (2014) deals with the mathematical derivation of effectiveness in the context of the adverse impacts of furlough programs on organizational productivity. Thereby, the following equation is used to compute effectiveness on a cost basis (Badiru, 2014):

$$ef = \frac{s_o}{c_o}, c_o > 0 \tag{6.1}$$

Where:

ef = measure of effectiveness on interval (0, 1)
s_o = level of satisfaction of the objective (rated on a scale of 0 to 1)
c_o = cost of achieving the objective (expressed in pertinent cost basis)

A completely achieved goal is expressed by the maximum satisfaction rating of 1, while a goal not achieved at all is 0. Considering the cost denominator, the goal achievement per unit cost, which can be money, time, or any other measurable resource, can be pertinently measured (Badiru, 2014).

Efficiency addresses resource allocation. It is efficient both to reach an expected goal with minimized resources, such as cost or time, and to maximize the result or output, respectively, from a given resource input. For computational purposes, we can leverage the mathematical formula for the extent of efficiency that is applied by Badiru (2014) as follows:

$$e = \frac{output}{input} = \frac{result}{effort} \tag{6.2}$$

Manzoor (2014) points out efficiency at all levels to be "one of the imperatives of public administration" and an inherent cultural part. Fiscal pressure is not the least that public goods and services providers, such as the armed forces, face continuously. Efficiency savings, such for overhead costs or bureaucracy depletion, can be reinvested in key capabilities. Once capacities are released, resource flexibility is enhanced. Resources can be expediently assigned to other projects, e.g., in the form of reorganizations in acquisition programs. In the Fiscal Year 2022, the operation and maintenance cost amounted to $319.2 billion, or 41.1% of the US Armed Forces Services budget (*Discretionary Budget Authority*), including day-to-day costs such as for training and operations, salaries of civilian staff, contract services for maintenance of equipment and facilities, fuel, supplies, and repair parts for weapons and equipment (DoD, 2022). In times of tight public funding, pervasive efficiency thinking and acting is required. In this regard, initiatives to reduce operating expenses, through energy conservation programs, reduction of procurement cost, cutting programs, or diligent reviews of staffing overheads, etc., are among the main drivers of efficiency.

Military services have faced up to this development and found various routes to foster efficiency initiatives. According to the Army's "*Climate Strategy,*" the US Army pursues energy efficiency as a means to accomplish its core mission (U.S. Department of the Army, 2022). The US Navy's "*Energy Program for Security and Independence*" (U.S. Department of the Navy, 2010) considers the value of energy as a critical resource and therefore envisions energy efficiency as a contribution to energy security, independence, improvement of operational effectiveness, and cost reduction.

The critical question about the conventional economic metrics is whether they are thoroughly well balanced in terms of resilience, robustness, and agility requirements. The measurement of effectiveness and efficiency allows an application of vital quantitative techniques in the decision-making process, but offers only a few anchor points to incorporate an evaluation of organizational benefits in a more comprehensive sense, tackling and including additional aspects such as:

- Strategic relevance, a systems view
- Economic resilience, operational agility, and organizational robustness

- Contribution to critical infrastructure safety and supply security
- Preparedness and adaptation capability toward disruptions, adversities, crises, and threats

The integration of the above-mentioned factors is part of the following models and analytical instruments. These tools are potentially suitable for closing the identified gap.

METHODOLOGY OF IMPLEMENTATION OF RESILIENCE METRICS

APPLICATION OF THE DEJI SYSTEMS MODEL®

If we presume a systems world and take a **system-thinking view** (Badiru and Agustiady, 2021) on the efforts to enhance resilience in the defense environment, we can confront and rectify the related implementation issues through the systematic involvement of additional technical-engineering and economic perspectives. **Systems engineering** addresses the integration of tools, people, and processes required to achieve a cost-effective, high-quality, and timely operation of the system. The characterization of the efficacy of this approach is handled, e.g., by Badiru, A. (2019) and by Badiru and Agustiady (2021). The nucleus of the application of systems engineering is a **system**. For reference, the typical definition of a system is revealed in the following (Badiru, 2019):

> A system is a collection of interrelated elements, whose collective output (together and in unison) exceeds the addition of the individual outputs of the elements.

The typical process for engineering problem-solving includes the following eight steps, which may be tweaked, condensed, or expanded depending on the specific analysis:

Step 1: Gather data and information pertinent to the problem
Step 2: Develop an explicit Problem Statement
Step 3: Identify what is known and unknown
Step 4: Specify assumptions and circumstances
Step 5: Develop schematic representations and drawings of inputs and outputs
Step 6: Perform engineering analysis using equations and models as applicable
Step 7: Compose a cogent articulation of the results
Step 8: Perform verification, presentation, and "selling" of the result

It is essential to have a systems tool that can be implemented to actualize the input-process-output framework. The **DEJI Systems Model®**, a trademarked systems tool for **system design, evaluation, justification**, and **integration** of any operational system (Badiru, 2012; Badiru, 2019), is designed to do this. Typical operational factors involved in a resilience build-up are covered in the model. Of particular importance within the application of the DEJI Systems Model is the integration stage.

Integration ensures that all elements that are essential for the system's operability are included in the process implementation phase. The DEJI Systems Model is effective for resilience-strengthening plans and initiatives. Its approach includes quantitative and qualitative elements, with the following four core stages:

1. Design
2. Evaluation
3. Justification
4. Integration

SYSTEM DESIGN

Design embodies any system initiative providing a starting point for a project and a definition of the end goal. The design stage represents a description of the requirements and specifications profile and can include technical product design or concept development, for instance. In terms of resilience, robustness, and agility, it is essential to incorporate corresponding parameters into the formulation of requirements and specifications of the respective initiative. Examples are armament acquisition programs or infrastructure projects requiring security peculiarities to tackle threats from cyberattacks, e.g., malware to manipulate the control of digital infrastructure. The imperative of the resilient system design is eventually based on the factor of damage prevention: each requirement should contribute to making the potential attack more difficult, elaborate, costly, and time-consuming, and less attractive for the perpetrator.

EVALUATION

Evaluation is the second stage of the structural application of the DEJI Systems Model®. The evaluation of a system pursuit can use a variety of qualitative and quantitative metrics. It encompasses the evaluation of the economic feasibility, beyond the technical sphere provided in the design stage. One of the pertinent techniques to enmesh resilience aspects into economic feasibility is the benefit–cost analysis, which is part of the instruments we advocate. It can be consistently integrated into this stage of the model and includes a quantitative view regarding the value of the robustness of systems, measures, programs, initiatives, projects, etc. A system platform may appear more cost-efficient. However, taking into account organizational benefits of resilience metrics, e.g., enhanced supply security, can significantly shift the advantage in viability.

JUSTIFICATION

Justification can be based on monetary, technical, or social reasons. The main focus in this stage of the DEJI Systems Model® is on the implementation and articulation of conclusions. Justification, in the context of the DEJI Systems Model®, is often more qualitative and conceptual than quantitative. Not all systems that are well designed and favorably evaluated are justified for implementation. Questions related to systems justification may include, e.g., the following:

- Desirability of the system for the operating environment
- Acceptability of the system by those who will be charged to run the system
- Sustainability potential for the new system

This means that the DEJI Systems Model® explicitly requires organizations to justify why a new system, initiative, program, endeavor, or approach is needed, particularly from a system-of-systems perspective. In this stage of the model, resilience, robustness, and agility factors have to pass a justification phase, which requires a critical qualitative assessment of whether a system pursuit sustainably meets the prevailing system environment's acceptability and desirability.

INTEGRATION

Integration is the last stage of the application of the DEJI Systems Model® and definitely the most critical part of any systems implementation. This is why operational practicality represents a matter of critical importance at this stage. A system that is not appropriately anchored to the prevailing operating environment may be doomed to failure. The DEJI Model explicitly requires that any system of interest be *integrated* into its point and manner of use. Integration needs to be done with respect to the standard operations within the organizations applying resilience aspects. For resilience reinforcement, the implementation requires consideration of the organization's operating mechanisms, e.g., security databases, training and education instructions, safety standards for vulnerable assets or confidential data, or crises and accident prevention and mitigation plans.

DEJI Systems Model® can positively impact how new systems are brought online in the prevailing operating environment, considering the existing tools, techniques, processes, and workforce within the organization. What is the value of integration of system characteristics over time? In this respect, an integral of the following form may be suitable for a mathematical exposition:

$$I = \int_{t_1}^{t_2} f(q)dq, \tag{6.3}$$

Where I = integrated value of quality, $f(q)$ = functional definition of quality, t_1 = initial time, and t_2 = final time within the planning horizon. An illustrative example is provided by Badiru (2023) for the case of a geometric alignment of the hypothetical physical parts of a system-of-systems.

QUANTITATIVE METHODOLOGY OF SYSTEMS MODELING

A quantitative technique that can be used for systems related to resilience reinforcement modeling is drawn from the general systems value model presented by Badiru

(2019). The model provides an analytical assessment of the value or impact caused by system components. A component can add to (+) or subtract from (–) from the overall health (or status) of the prevailing system. In this case, a system is represented as a p-dimensional vector:

$$V = f\left(A_1, A_2, ..., A_p\right) \tag{6.4}$$

where $A = \left(A_1, ..., A_n\right)$ is a vector of quantitative measures of tangible and intangible attributes. Examples of infrastructure **attributes** could be structural integrity, aesthetics, durability, resilience, capacity, adaptability, modularity, sustainability, reliability, and affordability. Attributes are considered to be a combined function of factors, x_1, expressed as:

$$A_k\left(x_1, x_2, ..., x_{m_k}\right) = \sum_{i=1}^{m_k} f_i\left(x_i\right) \tag{6.5}$$

where $\{x_i\}$ = set of m factors associated with attribute $A_k\left(k = 1, 2, ..., p\right)$ and f_i = contribution function of factor x_i to attribute A_k. Examples of **factors** could include flexibility, comfortability, utilization level, safety, and design functionality. Factors are themselves considered to be composed of indicators, v_i, expressed as

$$x_i\left(v_1, v_2, ..., v_n\right) = \sum_{j=1}^{n} z_i\left(v_i\right) \tag{6.6}$$

where $\{v_j\}$ = set of n indicators associated with factor $x_i\left(i = 1, 2, ..., m\right)$ and z_j = scaling function for each indicator variable v_j. Examples of **indicators** could be visual appeal, maintainability, acceptability, accessibility, escapability, modularity, and reconfigurability. By combining these definitions, a composite measure of the value of a process can be modeled as:

$$V = f\left(A_1, A_2, ..., A_p\right) \tag{6.7}$$

where m and n may assume different values for each attribute. A subjective measure to indicate the utility of the decision-maker may be included in the model by using an attribute weighting factor, w_i, to obtain a weighted PV:

$$PV_w = f\left(w_1 A_1, w_2 A_2, ..., w_p A_p\right) \tag{6.8}$$

where

$$\sum_{k=1}^{p} w_k = 1, \quad \left(0 \leq w_k \leq 1\right) \tag{6.9}$$

With this modeling approach, a set of resilience-related compositions can be modeled for the purpose of developing forecasts and projections about responses to disruptions and adversities.

To illustrate the model, suppose three acquisition options for armament equipment with

A: new system development
B: upgrade modification of an existing system
C: purchase of a commercially available technology

are to be evaluated based on four attribute elements: *capability, resilience, performance, modularity.* For this example, based on the equations, the value vector is defined as:

$$V = f\left(capability, resilience, performance, modularity\right) \quad (6.10)$$

Capability: "Capability" in this context, refers to the ability of the option to satisfy multiple and complex requirements. While a certain option may merely provide a basic configuration, a different option may be capable of meeting additional functionalities. In the study, the levels of increase in capability of the three acquisition options are 45%, 36%, and 28%, respectively.

Resilience: "Resilience" represents the ability of the armament system to withstand and recover rapidly from disruption, particularly from attacks, accidents, and malfunctions. In the analysis, the resilience levels of the three competing options are valued at 18%, 27%, and 14%, respectively.

Performance: "Performance" refers to the ability to satisfy schedule and cost requirements. In the example, the three options can, respectively, satisfy requirements on 20%, 34%, and 46%.

Modularity: "Modularity" can be measured by the degree of contribution of a separate option to the functionality and quality of the complete whole of the system-of-systems when combined with other armament parts. For the example, the three options, respectively, show a modularity level of 22%, 38%, and 42%.

Option A is the best alternative in terms of capability measure but scores poorly in terms of the performance metric. Option B shows the best resilience rating and also a high modularity measurement. Option C provides the best performance and modularity score, but nonetheless, both its capability and resilience measurement are comparatively low to justify the integration into an environment with expected preparedness and resistance toward adversities. Subject to an even distribution of scores to all four vector attributes, Option B reaches the slightly best overall rating.

The analytical process can include a lower control limit in the quantitative assessment, such that any option providing value below that point, e.g., for resilience, is not acceptable and results in exclusion. Similarly, a minimum value target can be incorporated into the graphical representation, such that each option is expected to exceed the target point on the value scale.

LIFE CYCLE RESILIENCE ASPECTS APPROACH

A common economic instrument to consider costs of different options, programs, measures, etc., comprehensively is the life cycle cost approach: investment and operating costs along the respective life cycle from creation to termination, including, e.g., initial and research cost, cost for staff, infrastructure, maintenance, and other services, taxes, fees, training, exercises, travel, etc.

We suggest converting this instrument into a **life cycle resilience approach** that consistently incorporates measures to achieve preparedness toward vulnerabilities, mitigate possible incidents of adversity, respond to actual ones, and to recover rapidly from disruptions, crises, and catastrophes. The US Army's *"Climate Strategy"* reflects an approach in this regard aiming at climate resilience and sustainability, especially to conserve resources and reduce operating costs (U.S. Department of the Army, 2022), an endeavor we can pertinently amend for the purpose of a comprehensive resilience assessment. The Army's Leadership in Energy and Environmental Design certification in the US Green Building Council implicates extensive standards for sustainable and efficient infrastructure, such as for new building constructions or major renovations, covering all aspects of construction and operation (U.S. Department of the Army, 2022). This context offers an inspirational insight for resilience build-up, e.g., in the form of a resilience certification system for the inclusion of relevant aspects in the life cycle of projects.

Figure 6.1 illustrates the range of life cycle–related categories, factors, and resources that appear useful to certify contribution to the resilience reinforcement of organizations in various endeavors. In this regard, there are four life cycle domains in terms of the consideration of resilience enhancement aspects: prevention, readiness, management, and wrap-up (BMI, 2022). They are necessarily based on a well-balanced and comprehensive consideration of the resource perspective, an aspect that is also depicted in the figure.

MANAGEMENT OF RESILIENCE PROGRAMS

An advanced management concept may incorporate a system of targets and key figures based on the performance measurement principle. Metrics can be installed and monitored on every level of the above-introduced resilience life cycle, from prevention to wrap-up. The advantage of such an extension and operationalization, respectively, lies in its support of the management information system of each organization, since the multifaceted approach to boosting resilience can be divided into smaller measurable, and therefore adequately manageable, chunks. The following are examples of applicable metrics in terms of a target and key figure system:

- Number of awareness training participants and quality improvement of test results

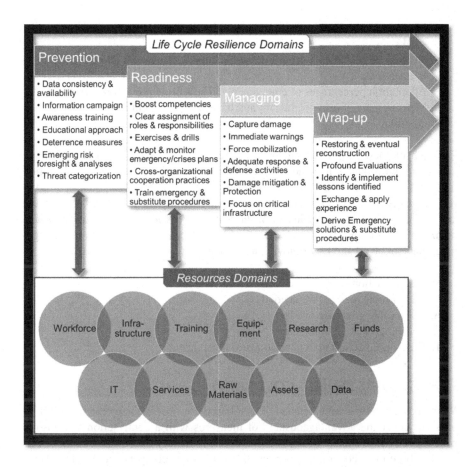

FIGURE 6.1 Life Cycle Resilience Aspects Approach (Adapted from German Federal Ministry of the Interior and Community (BMI), German strategy for strengthening resilience to disasters. Implementation of the Sendai Framework for Disaster Risk Reduction (2015–2030) – Germany's contribution 2022–2030, Berlin, July 13, 2022. https://www.bundesr-egierung.de/breg-de/service/publikationen/deutsche-strategie-zur-staerkung-der-resilienz-gegenueber-katastrophen-2062054)

- Number of introduced and successfully completed resilience education programs
- Reachability percentage of organization members in case of emergency notifications
- Participation of the workforce in exercises or drills
- Level-based measurement of successful infrastructure conversion projects
- Number and volume of actually pertinent research initiatives linked to resilience
- Energy conservation outcomes and achievements

- Measurement of the logistical capacity for the build-up of inventories and stocks
- Number of acquisition programs with a consideration of resilience metrics in assessments
- Adaptation percentages of emergency and substitute procedures trainings
- Measurement of improvement of available data, pertinent data quantity, and quality

For each of these examples and many more, concrete goals can be set, specific key figures introduced, and the progress monitored to derive necessary adaptation measures. Resilience build-up commitments thereby can be operationalized in the respective organization. Of utmost importance, regarding fully applicable target and key figure systems, is the inclusion of the **SMART principle of operations management**. Refer to the following list for the SMART principle of operations management that can be adapted for innovation management:

S: Specific actions are needed
M: Measurable metrics are needed
A: Achievable goals are needed
R: Relevant plans are mandatory
T: Time-based action plan is of utmost importance

QUANTITATIVE BENEFIT–COST RATIO

We moreover endorse an extension of cost-based considerations by the factor "**benefits**" to cover the advantageousness of measures with a contribution to boosting resilience, agility, and robustness, expressed in monetary units. The beauty of this tool lies both in its economic representability and in its compatibility with the DEJI Systems Model®, particularly in the evaluation stage.

The **benefit–cost ratio**, as applied by Badiru and Omitaomu (2011) or Cellini and Kee (2015), for instance, forms the underlying principle. This model relates the benefits of a certain object of consideration, e.g., a certain program, initiative, or investment option within a feasibility study, to its costs. It is a cash flow–based approach taking into account the monetized value of the benefits, on the one hand, and the present value of the costs within a certain time frame, on the other. The benefit–cost ratio (BCR) is thereby expressed by the following fraction (Badiru and Omitaomu, 2011):

$$BCR = \frac{Present\,Value\,of\,aggregated\,Benefits}{Present\,Value\,of\,aggregated\,Costs} = \frac{\sum_{n=1}^{T} B_n (1+r)^{-n}}{\sum_{n=1}^{T} C_n (1+r)^{-n}} \quad (6.11)$$

where

BCR represents the benefit–cost ratio of the particular project, policy, program, etc.

C_n comprises the costs of the initiative, which consist of the projected annual net cash outflows (disbursements) for the respective year n within the period of consideration T

B_n covers the monetary value of the total annual benefits for the respective year n within T

n is the year of the annual net cash outflows or of the total annual benefits

r is the discount rate

T is the period of consideration of the option (project, policy, program, etc.) in years

The present worth of the aggregated benefits has to exceed the present worth of costs to justify the respective acceptability – otherwise, the initiative is not beneficial. In mathematical terms, this cognition means the ratio has to exceed 1 (Badiru and Omitaomu, 2011). The ratio can be increased either by enhancing the nominator or by diminishing the denominator, expressed as follows:

$$BCR \uparrow = \frac{Benefits \uparrow}{Costs} \quad or \quad BCR \uparrow = \frac{Benefits}{Costs \downarrow} \tag{6.12}$$

The BCR is a fertile tool to compare different options. The advantageousness of an option is reached by comparison. The option with the highest ratio between the present value of the project, policy, endeavor, or program benefits and the present value of the aggregated disbursements is beneficial, whereas the value has to be above 1 to be regarded in the first place.

The BCR is based on **quantitative** measurement. Cost forecasting is an immanent part of most endeavors anyway. For the benefit–cost methodology, the project, initiative, program, policy, or option needs additionally to be analyzed on the monetary value of benefits, which can include the monetized utility contribution or avoided costs. A major prerequisite is to consider the requirement of converting identified benefits into a **monetary value** based on comparable units (Williams and Phillips, 2011). There are abundant scholarly applications of the benefit–cost analysis in the quantitative dimension, e.g., by Boness and Schwartz (1969) on the choice of a replacement policy for specific military aircraft in a specific assignment, or by Byrns et al. (1995) presenting a merit function for evaluation of complex military systems. The following are general examples of benefits that are translatable to quantitative parameters in the form of monetary units:

- Reduction of lead/turnaround/processing times
- Diminution of scrap and error rates
- Improvement of delivery readiness and adherence to delivery dates
- Improvement of work processes and working conditions
- Amplification of skills, knowledge, and capabilities of the workforce
- Improvement of service or product quality and reduction of penalty costs related to quality
- Enhancement of customer satisfaction and reduction of customer complaints

- Improvement of employee motivation and increase of attractiveness to talents
- Improvement of the organization's image and reputation
- Reduction of overtime and avoidance of extra work
- Diminution of energy consumption and increase in energy efficiency

In the light of resilience-linked initiatives, programs, policies, projects, or options, an adjustment of the quantitative measurement of benefits necessarily has to be undertaken in order to apply the methodology accordingly specified. We suggest leveraging the following domains of benefits toward a pertinent application of quantitative metrics (Figure 6.2). The resilience-related perspective can be thereby enmeshed and evaluated within the microsphere of an organization.

Historical data on the devastating effects of hybrid threat scenarios, e.g., related to cyberattacks or physical damage of critical infrastructure, is a useful enabler to estimate the quantitative extent of resilience-linked benefits. The crucial question and at the same time, a break-even scenario – an option, a project, a policy, or program with a BCR that exceeds the value of 1 proves to surpass this point – is whether the costs of potential damage, loss, or impairment that could be specifically prevented, mitigated, or protected against by measures of resilience build-up undercut the benefits from resilience efforts. This is a question of strategic significance, since menaces to the workforce, infrastructure, information technology (IT) systems, assets, etc. are strongly mission critical.

For instance, a new system of storage facilities for military equipment with redundancies in its dislocation triggers investment and operating costs. The two main alignment opportunities include specialization versus redundancy. An oversized facility may offer synergies from a cost point of view. The facility could be specialized in a certain system, branch of service, etc. However, from a hybrid threat perspective, disruption and adversity augment substantial risks to operational readiness in terms of defense capabilities. This risk can be tackled by redundancy, which impairs cost efficiency. The benefit–cost methodology provides a quantitative instrument to

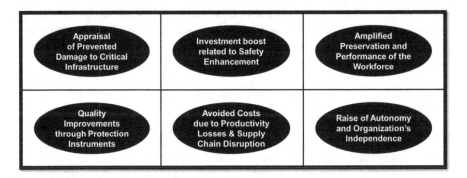

FIGURE 6.2 Examples of Measurable Quantitative Domains of Benefits (Benefit–Cost Ratio)

tackle the presented issue, since the benefits from redundancy can be interrelated to the additional costs.

VIABILITY EVALUATION WITH COMPLEMENTARY QUALITATIVE BENEFIT ANALYSIS

There is an alternative approach to handling benefits. As a case example, we suppose a defense cost analysis agency conducts a viability study project on the follow-up organizational solution for the provider of military clothing and personal equipment on behalf of the armed forces. According to the requirements, the demanded services include the provision, exchange, return, conversion, reprocessing, and repair of clothing and personal equipment as well as the overall management of the corresponding processes and ensuring of the service levels. We assume the service provider is not necessarily part of the military organizational structure, as providing capability of services for products of this category are not an obligatory part of the defense core competencies. The cost analysis agency examines several options, from a full in-house model, over different mixed and cooperation options, to a full outsourcing solution. Outsourcing does not meet the requirements of full supply security, which is of strategic relevance for the functional capability in terms of operational readiness of the military branches of service. This option already drops out in the first examination stage of the economic feasibility evaluation, as the chance of service disruption does not meet the supply security requirements in the absence of opportunities for government influence. Subsequently, a full cost computation of the investment and operating costs for the four remaining options is performed – including a new government agency, a private in-house company, contract-based cooperation with the private enterprise, and a new service-providing corporation with state and private joint shareholding.

The monetary evaluation is based on the **capital value method** (Vanek et al., 2016). Therefore, the **net present values** of the four options on the basis of their future cash flows are calculated and made comparable. All expected cash inflows and inflows of the whole period of consideration are discounted to the same reference date in the year the decision on the follow-up solution is made.

The service provider in each is supposed to render services for the customer "armed forces" and is thereby accordingly paid by public funds, but however, providers differ in the respective organizational costs due to different resource use models. Since it can be assumed that the expenses will exceed the revenues, the option with the lowest negative present value is economically beneficial in this stage of the viability approach. The following equation is used to compute the net present values:

$$NPV = \sum_{n=1}^{T} \frac{CF_n}{\left(1+r\right)^n} \tag{6.13}$$

where

> *NPV* describes the net present value of the respective organizational option
> *CF* is the annual net cash flow forecast for each year of the period of consideration (defined by cash inflows – cash outflows in the respective year)
> *n* is the year of the annual cash flows
> *r* is the discount rate
> *T* is the period of consideration of the viability study project (in years)

Typically, monetary evaluation is enriched by a sensitivity analysis. In addition to this, risk analyses are another classic component of quantitative economic feasibility. Thereby, the monetary value of risks related to the respective option is assessed and incorporated into the consideration on a present value basis. Monetarily quantifiable risks are assessed by identifying risks tailed to the respective project and considering the probability of occurrence as well as the forecast damage level.

A risk matrix as depicted in Figure 6.3 can help to classify the risk categories and to quantify the impact for the purpose of integration into the viability study.

Within the framework of quantitative economic feasibility studies, risk analyses offer the tools to incorporate resilience, robustness, and agility factors. Statistically validated data, for instance, for infrastructure damage, cyberattacks, construction or delivery delay, supply chain disruption, or lack of skilled labor force or other resources, can be transferred into monetary risk surcharges. This approach helps to provide transparency for financial decisions. In addition, the aspect of resilience is methodically enmeshed – risk assessment is an immanent part of economic viability.

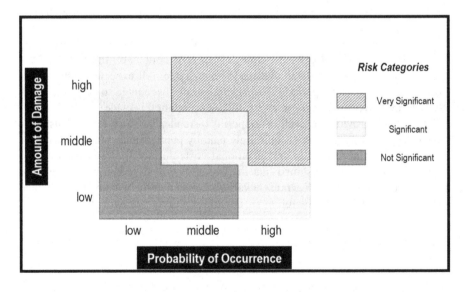

FIGURE 6.3 Example of a Risk Matrix as Part of a Risk Analysis

The quantified risks are to be discounted with the annual cash flows. They have a modifying effect on the previously calculated net present values of the options. According to the present values, the consecutive order of the advantageousness of options in terms of viability based on the capital value method is impacted by the risk consideration.

We consider the calculation of the present values to be the quantitative component of a bipolar assessment. As a further complementary part of the decision-making process, we suggest a **non-monetary evaluation** of the benefits from certain qualitative criteria with relevance for the contribution to mission success. They can be incorporated into the above-mentioned viability study as a separate element, based on the **utility** profile of each option. The techniques of utility modeling can be used to assess the relative utilities of sub-elements in the system (Badiru and Foote, 1992).

The non-monetary assessment methodology is based on comparisons of evaluation points. Applied on the project to referee the follow-up organizational solution for the future provider of military clothing and personal equipment, therefore, the **qualitative benefit analysis** addresses the degree to which the utility value of different organizational features is achieved. First, the weighting of the used evaluation criteria has to be specified. In our example, we overall determine six qualitative organizational characteristics as criteria. The criteria used to evaluate the measure, endeavor, project, etc. are to be determined on a project-specific basis as well as weighted according to their importance (BMF, 2011). In order to prevent multiple crediting, the individual evaluation criteria should not overlap. It is recommended to avoid contradictive criteria. In our example, the feature "resilience" is supposed to carry the highest importance and is weighted accordingly. Subsequently, the evaluation point scale of the benefit analysis is to be defined.

We set a 1 (minimum) to 10 (maximum) point scale, where the value of 0 would lead to the exclusion of an option. An appropriate evaluation would also include details on the assessment scale, i.e., a precise description of the degree of achievement that is required to meet a certain score within the assigned scale. To calculate the overall benefits per criterion (**partial utility value**), the weighting of each organizational feature is multiplied by the achieved score, respectively. The utility value of an option is obtained by addition of the partial benefits. The chosen characteristics to assess the utility are Agility, Governance, Resilience, Modularity, Performance, and Capability.

Agility encompasses the ability of an organization to renew and change quickly, and to succeed in a rapidly changing, ambiguous, turbulent environment or under new conditions (McKinsey, 2015).

Governance refers to the scope of effective control, supervision, management, and regulatory opportunities for the superior agency or command regarding the responsibilities within the defense organization structure.

Resilience represents the ability of the organization to withstand and recover rapidly from disruption and adversities. It includes the ability to cope with crises or attacks to return to a normal state and develop mechanisms to be ready against future similar attacks (Desouza and Xie, 2021).

Modularity includes the degree of contribution of a separate organization to the functionality and quality of the complete whole of the defense system-of-systems when combined with other organizations and functions.

Performance embraces the organization's ability to meet the schedule and cost requirements and includes the ability of effective resource acquisition.

Capability refers to the ability of the organization to satisfy multiple and complex requirements.

Mertens and Badiru (2023) present an hypothetical result of a qualitative benefit analysis for the applied case considering the above-mentioned features, point scale, and principles of weighted partial utilities. Each organization feature has a certain weighting factor due to its importance (brackets). The scale of 1 to 10 multiplied with this factor yields the partial utility. The sum is the total utility per option. In comparison, one option obtains the highest score, not least due to the advanced satisfaction of the strongly-weighted "resilience" criteria. However, according to the quantitative analysis of present values, Option D had the strongest advantageousness. How can we combine these results? Mertens and Badiru (2023) present further details about this.

QUALITY–COST TRADE-OFF

Organizational robustness, resilience, and agility aspects have been adequately enmeshed into the "viability study with complementary qualitative benefit analysis" approach. The incorporation has been made in the risk analysis as well as in the qualitative utility criteria. The quantitative and qualitative assessments result in a decision advice. In the decision-making sphere, all the different factors have to be pulled together to form a consistent, transparent, and comprehensible proposal.

We can support this last part of the study endeavor methodically by providing a means to relate the quantitative and qualitative results appropriately. We endorse a 70%–30% partition with linear interpolation. Hence, the decision is prepared to be weighted 70% by costs – the present value of the options considering their risks – and 30% by utility – the score of the benefit analysis.

The following sequence addresses the trade-off between quality and cost that the decision process has to confront and rectify in this context:

1. Apply linear interpolation on the present values of the options (normalized 100 scale basis)
2. Scale the utility values of the options on a normalized 0 to 100 basis (benefit score)
3. Weight the cost and the quality dimension with regard to the 70%–30% partition
4. Scale the score by the factor 100, and rank the results based on the highest achieved score

1. The linear interpolation is achieved with the following equation:

$$CS_i = \frac{x*C_{min} - C_i}{C_{min}} * S_{max}, \quad with \ C_i \leq x*C_{min} \tag{6.14}$$

where

CS_i is the cost score that a certain option i obtains according to its cost C_i
x is the factor for the linear interpolation
S_{max} is the maximum cost score of 100 achievable for the lowest cost
C_i is the cost of the option i, expressed by its present value
C_{min} is the cost of the option with the lowest cost, expressed by its present value

Note: In the quantitative part of the study, we deal with negative present values, since the cash outflows are assumed to exceed the cash inflows by far. For the applied cost dimension, their sign is to be changed correspondingly. In general, the methodology of the lowest cost is to be applied accordingly.

The option with the lowest cost – applied to the study with the lowest negative present value – obtains a cost score of S_{max} = 100, which is the first supporting point of the linear interpolation. An option that has x times higher cost or above comes to a cost score of 0, the second supporting point of the linear interpolation. We use $x = 2$ in our case example, i.e., costs twice as high as C_{min} or above yield a cost score of 0.

2. The partial utility values of the respective options are based on a 0 to 100 scale and have to be modified in terms of a comparable cost–benefit scale. Hence, we use the normalized 0 to 100 basis.
3. To objectify the cost–quality trade-off, it appears advisable to use weighting factors for the quantitative and the qualitative dimension to prepare balanced decision-making. We advocate the decision sphere to be based on a 70% (cost) to 30% (utility) relation. Therefore, the weighting factors ω_C = 0.7 (cost dimension) and ω_U = 0.3 (quality dimension) are applied on the scores.
4. Finally, in order to consider the decimal places appropriately, we scale the total score after weighting by the factor 100 and rank the results based on the highest achieved score.

Although option D had the lowest negative present value and appeared beneficial in light of the cost dimension, option B has the highest overall score considering the results of the qualitative assessment. Even the weighting factor of 30% does not impede the advantageousness of option B, not least due to its comparative qualitative advantage in the evaluation of organizational resilience.

It is advisable to add a third important parameter to the considered dimensions of quality and cost, namely, time. This dimension can be similarly incorporated into the decision-making process and allows an overall more comprehensive procedure. The approach of linear interpolation offers the means to compare options fulfillment levels in this dimension, e.g., including the research and development phase. The weighting factors partition has therefore to consider a third dimension. **Quality, time**, and **cost** are the three momentous parameters, the nucleus, to assess the feasibility as well as the efficiency of a financial decision to be carried out from a **Quality–Time–Cost Trade-off Investment Space**. There are multifarious inter-relations between these three investment parameters. They are traditionally all necessary in an investment decision setting.

AREAS OF ACTION TO INCLUDE ROBUSTNESS, AGILITY, AND RESILIENCE

Beyond the common effectiveness and efficiency metrics, we have demonstrated various quantitative and qualitative tools to incorporate resilience requirements in defense activities and measures. In the course of this, the thematic range of initiatives, programs, and policies that bear relevant elements to enmesh resilience angles is almost inexhaustible. We encourage readers to consider the aspects pervasively and hereinafter illustrate examples of pertinent areas of action.

- *Legislative measures:* According to NATO, civil preparedness is one of the central pillars of resilience and collective defense (NATO, 2022b). To ensure a cohesive society-wide readiness to disruption, disasters, crises, risks, and hybrid as well as armed threats, the statutory regulations, e.g. the emergency legislation, are to be kept up to date facing current and upcoming developments. Where necessary, the legislation should be amended.
- *Energy conservation:* Areas of civil society and the armed forces that successfully enhance efforts to conserve energy resources boost independence and flexibility, and also reduce vulnerability. Being less vulnerable strengthens overall resilience from the bottom up.
- *Diversification:* Every organization, including the armed forces, is supplied with goods or services. A diversified supply chain, in terms of both the product variety itself as well as the providers, contributes to fail-safety and resilience build-up. Reliability, sustainability, and supply security are paramount in this regard. Their importance for crisis preparedness and successful mission completion only increases with the strategic relevance of the products.
- *Fallback Options:* In Both civil and military preparedness toward adversities, disruptions, and crises, it is advisable to incorporate fallback, work-around, and contingency options in operations. In case of adverse fallouts in the main operational environment, the alternative solution is not only to be announced but also to be prepared and practiced.
- *Strengthen Core Competencies:* Risks and threats require strong state responses. Capable government, defense, and agency structures enable appropriate handling and mitigating of security threats. Therefore, robust organizational structures and processes have to remain up to an adequate

extent under state control. Outsourcing and efficiency consideration find limits in fields of strategic relevance. Without a strong focus on core competencies that necessarily have to remain under strategic control of the government, armed forces, and authorities, a coordinated chain of action in response to any sort of disruption is impaired.

- *Main Focus of Resilience:* Where vulnerabilities, adversities, and disruptions have the potential to cause the most severe damage to national security, particularly expressed by military operational readiness and civil society's ability to function, resilience efforts should be most focused. Examples are key technologies for armament as well as sectors of critical infrastructure. An additional example is the US measures in the field of cybersecurity, recently reflected by the *"Fiscal Year 2022 Cybersecurity Sprints,"* which are designed to operationalize the vision, to raise public awareness, and to drive action (DHS, 2022). A further example is President Biden's Executive Order 14057, which outlines a strategy to prepare US Federal agency operations in terms of adaptation and resilience plans against climate impacts (U.S. President, 2021).
- *Workforce agility:* Organizational resilience cannot be achieved without adequate consideration of the workforce sphere. The impacts of the Covid-19 pandemic on mission accomplishment, including in the armed forces, have demonstrated the necessity of agile instruments to keep staff on duty in times of volatility and crisis. As an example, in light of the U.S. Telework Enhancement Act (U.S. Government, 2010), the 2021 Guide to Telework and Remote Work in the Federal Government emphasizes that telework serves as an instrument of a flexible workplace and to improve labor force performance and engagement, among others (U.S. Office of Personnel Management, 2021). Telework is proposed to leverage the conduction of mission-critical services through a safe environment in an emergency situation (U.S. Office of Personnel Management, 2021). There are other resilience factors linked to the labor force and job criticality that are expected to influence the capability of a flexible accomplishment of system-critical work preparedness to disruptions, adversities, vulnerabilities, and crises. Examples are enhancing and maintaining the qualification and education levels for areas of complex tasks, diminishing vacancies, and conducting awareness training or drills.
- *Redundancy and Reserves:* Resilient structures, processes, and procedures in terms of defense operability and civil preparedness are preponderantly related to the availability of reserves. Reserves include labor force as well as supply of goods, for instance, spare or replacement parts for weapon systems, disposables, ammunition, or personal equipment. A supply system for these goods that solely relies on specialization is likely to fail in case of disruption. Appropriate redundancy requirements are a necessary expansion of the key principles in the functionality parameters of such system domains.

- *Reservation of capacities for crises:* Certain services, e.g., transport capacities on land, at sea, or in the air and space, require lead times for realization. Sustainable and crisis-proof armament supply requires continuous utilization of civil production facilities and preparedness toward adjustments. Hence, reservation thinking should be included, for instance, in feasibility studies or contracting with respect to the quality–time–cost trade-off.
- *Acquisition process*: The acquisition process for defense products offers a variety of opportunities to incorporate resilience requirements into procurement evaluation metrics.
- *Education and training:* There are numerous entry points in the education and training efforts of defense organizations to consider resilience, robustness, and agility comprehensively. Examples are training certificates, research alignment, or the skill and qualification levels to sustain crisis-resilient, prepared staff for complex demands.

CONCLUDING DISCUSSIONS

Resilient structures, organizations, and processes are intended to ensure preparedness as well as to maintain operability even during any hiccup, adversity, or disruption – any unforeseen events, any interruption, any crisis or shock, any subtle risk or actual threat, either domestically or elsewhere, anywhere around the world. The study is designed to raise awareness of decision-makers toward the diversity of robustness, resilience, and agility factors involved in defense services, operability, and capabilities as well as civil preparedness toward disruption, adversities, crises, threats, and risks in light of national security. We thereby anticipate that this conceptual framework will foster adequate enmeshment of resilience criteria in quantitative approaches and analytical tools and techniques, respectively, in the defense context. The purpose of this chapter also involves a contribution to the resilience debate in terms of critical infrastructure safety, military capabilities, societal sensitization, supply security, and government commitment toward resilience build-up in societies in general. Additionally, we aim to demonstrate multifaceted problem-solving applicability and to spark interest in the DEJI Systems Model®, a life cycle view on resilience aspects, and the wide range of quantitative and qualitative elements of cost–benefit analyses.

We derive the following findings from our examinations. The traditional economic metrics of effectiveness and efficiency have too short a reach to fully cover the aspect of resilience build-up in defense activities. We endorse suitable supplementary instruments. The DEJI Systems Model® as well as the presented quantitative techniques of systems modeling apply to resilience-strengthening initiatives. With a system-of-systems view, we make sure the right outputs can be integrated into the prevailing operative environment of defense organizations. The DEJI Systems Model® offers the platform for a multidimensional analysis of the resilience factors, considering many of the typical *"ilities"* attributes related to a systems view. The model focuses on the specific goals of a system, considering the specifications, prevailing constraints, possible behaviors, and structure of the system, and involves a consideration of the activities required to ensure that the system's performance matches specified goals. Concretely, the measures and activities to meet resilience objectives pass through a system engineering perspective and its particular integrative process mythology in a stage-by-stage approach. The model can be used to evaluate and justify a designed measure, with a specific template for integration into the expectations and peculiarities of the prevailing system. We adapted life cycle costing to a life cycle resilience consideration approach and thereby demonstrated comprehensiveness from this point of view. It is anticipated that this technique will be suitable for strategic documents and national proposals linked with organizational resilience due to the wide range of categories, factors, and resources that are included. The management of programs that arise from the strategic documents and national proposals with a linkage to resilience reinforcement can be supported by integrated systems of targets and key figures, based on the SMART principle of operations management. To enmesh utility as an additional factor to assess viability, we examined two methodologies: the quantitative benefit–cost ratio as well as the viability evaluation with complementary qualitative benefit analysis. While the benefit–cost ratio encompasses the monetary value of benefits and therefore facilitates the assessment of individual programs, initiatives, measures, projects, or activities, the qualitative benefit analysis is tailored to a viability study scenario, where options are compared on the non-monetary value of their utility contribution as a quality factor in addition to the cost dimension in decision-making with respect to the quality–cost trade-off. Both techniques offer beneficial tools to incorporate resilience-related requirements and assess their value in the light of overall feasibility. In addition to that, we presented miscellaneous strategic fields that enable moving forward in resilience-enhancing activities. These thematic fields offer connection points to consider the models we proposed and applied in this chapter. We advocate future research efforts that utilize live data of actual projects, programs, and initiatives, and demonstrate the practical implications of our conceptual framework, particularly in the defense context.

REFERENCES

Badiru, A. B. and T. Agustiady (2021). *Sustainability: A Systems Engineering Approach to the Global Grand Challenge*, Taylor & Francis CRC Press, Boca Raton, FL.

Badiru, A. B. (2012). Application of the DEJI Model for Aerospace Product Integration. *Journal of Aviation and Aerospace Perspectives (JAAP)*, Vol. 2, No. 2, Fall, pp. 20–34.

Badiru, A. B. (2014). Adverse Impacts of Furlough Programs on Employee Work Rate and Organizational Productivity. *Defense Acquisition Research Journal: A Publication of the Defense Acquisition University*, Vol. 21, No. 2, pp. 595–624.

Badiru, A. B. (2019). *Systems Engineering Models: Theory, Methods, and Applications*, Taylor & Francis/CRC Press, Boca Raton, FL.

Badiru, A. B. (2023). *Project Management for Scholarly Researchers: Systems, Innovation, and Technologies*, Taylor & Francis CRC Press, Boca Raton, FL.

Badiru, A. B. and B. L. Foote (1992). Utility Based Justification of Advanced Manufacturing Technology. In H. Parsaei, W. Sullivan, and T. Hanley, Eds. *Manufacturing Research and Technology*, Vol. 14, Elsevier Science Publishers, New York, pp. 189–207.

Badiru, A. B. and O. Omitaomu (2011). *Handbook of Industrial Engineering Equations, vFormulas, and Calculations*, CRC Press, Taylor & Francis Group, Boca Raton, FL, ISBN: 978-1-4200-7627-1

Boness, A. J. and A. N. Schwartz (1969). A Cost-Benefit Analysis of Military Aircraft Replacement Policies. *Naval Research Logistics Quarterly*, Vol. 16, No. 2, pp. 237–257. https://doi.org/10.1002/nav.3800160208

Byrns Jr, E., J. Corban, and S. Ingalls (1995). *A Novel Cost-Benefit Analysis for Evaluation of Complex Military Systems*, Military Academy West Point, West Point, NY.

Cellini, S., and J. Kee (2015). Cost-Effectiveness and Cost-Benefit Analysis. In K. Newcomer H. Hatry, and J. Wholey, Eds. *Handbook of Practical Program Evaluation*, Vol. 636, John Wiley & Sons, Inc., Hoboken, NJ.

Cîrdei, I. A. (2018). Improving the Level of Critical Infrastructure Protection by Developing Resilience. *Land Forces Academy Review*, Vol. 23, No. 4, pp. 237–243.

CNBC. (2022). All You Need to Know about the Nord Stream Gas Leaks — And Why Europe Suspects 'Gross Sabotage'. Published/Updated: October 1, 2022. Written by: Meredith, S. https://www.cnbc.com/2022/10/11/nord-stream-gas-leaks-what-happened -and-why-europe-suspects-sabotage.html (accessed December 2, 2022).

Cybersecurity, and Infrastructure Security Agency (CISA). (2022). Infrastructure Security: Critical Infrastructure Sectors. https://www.cisa.gov/critical-infrastructure-sectors (accessed November 10, 2022).

Desouza, K. and Y. Xie (2021). Organizational Robustness and Information Systems. In *Proceedings of the 54th Hawaii International Conference on System Sciences*, pp. 6089–6098.

DoD. (2020, July 2). DoD Directive 2311.01 – Law of War Program. *Originating Component: Office of the General Counsel of the Department of Defense. Effective.*

German Federal Ministry of Finance (BMF). (2011). Arbeitsanleitung Einführung in Wirtschaftlichkeitsuntersuchungen (Work Instruction – Introduction to Economic Efficiency Studies). GZ: II A 3 - H 1012-10/08/10004. Circular letter of the Federal Ministry of Finance of January 12, 2011, changed by Circular letter from May 7, 2021 – II A 3 - H 1012-6/19/10003:003, DOK 2021/0524501. https://www.verwaltungsvors chriften-im-internet.de/bsvwvbund_20122013_IIA3H1012100810004.htm (accessed December 19, 2022).

German Federal Ministry of the Interior and Community (BMI). (2022). Deutsche Strategie zur Stärkung der Resilienz gegenüber Katastrophen. Umsetzung des Sendai Rahmenwerks für Katastrophenvorsorge (2015–2030) – Der Beitrag Deutschlands 2022–2030 (German strategy for strengthening resilience to disasters. Implementation of the Sendai Framework for Disaster Risk Reduction (2015-2030) - Germany's contribution 2022-2030). Berlin, July 13, 2022. Serial Number: BMI22017 https://www .bundesregierung.de/breg-de/service/publikationen/deutsche-strategie-zur-staerkung -der-resilienz-gegenueber-katastrophen-2062054 (accessed December 8, 2022).

Harraf, A., I. Wanasika, K. Tate, and K. Talbott (2015). Organizational Agility. *Journal of Applied Business Research (JABR)*, Vol. 31, No. 2, pp. 675–686. https://doi.org/10 .19030/jabr.v31i2.9160

Linkov, I., B. D. Trump, K. Poinsatte-Jones, P. Love, W. Hynes, and G. Ramos (2018). Resilience at OECD: Current State and Future Directions. *IEEE Engineering Management Review*, Vol. 46, No. 4, pp. 128–135. https://doi.org/10.1109/EMR.2018 .2878006

Manzoor, A. (2014). A Look at Efficiency in Public Administration: Past and Future. *SAGE Open*, Vol. 4, No. 4. https://doi.org/10.1177/2158244014564936

McKinsey. (2015, December 1). The Keys to Organizational Agility. Aaron De Smet – Defining Organizational Agility. https://www.mckinsey.com/capabilities/people-and -organizational-performance/our-insights/the-keys-to-organizational-agility (accessed November 30, 2022).

Mertens, A. and A. Badiru (2023). Compatibility of Qualitative and Quantitative Resilience Requirements with Economic Metrics in the Defense Context, Special Study Report. Graduate School of Engineering and Management, Air Force Institute of Technology (AFIT), Wright Patterson Air Force Base, Dayton, OH.

New York Times (2022). Russian Attacks on Ukraine's Power Grid Are Endangering Nuclear Plants, a U.N. Agency Warns. Published: November 17, 2022. Written by: Santora, M. https://www.nytimes.com/2022/11/17/world/europe/russian-attacks-on-ukraines -power-grid-are-endangering-its-nuclear-plants-a-un-agency-warns.html (accessed December 1, 2022).

New York Times (2021). Cyberattack Forces a Shutdown of a Top U.S. Pipeline. Published: May 8, 2021. (updated: May 13, 2021). Written by: Sanger, D., Krauss, C., and Perlroth, N. https://www.nytimes.com/2021/05/08/us/politics/cyberattack-colonial-pipeline.html (accessed November 17, 2022).

North Atlantic Treaty Organization (NATO). (2022a). 2022 NATO Summit 28 June 2022– 30 June 2022. Last updated: July 01, 2022. https://www.nato.int/cps/en/natohq/news _196144.htm (accessed November 14, 2022).

NATO. (2022b). Resilience, Civil Preparedness and Article 3. Last updated: September 20, 2022. https://www.nato.int/cps/en/natohq/topics_132722.htm (accessed November 15, 2022).

President, U.S. (2021). Executive Order 14057: Catalyzing Clean Energy Industries and Jobs Through Federal Sustainability. December 8, 2021. *Federal Register*, Vol. 86, No. 236, pp. 70935–79943.

Ross, R., V. Pillitteri, R. Graubart, D. Bodeau, and R. McQuaid (2019). Developing Cyber Resilient Systems: A Systems Security Engineering Approach. (No. NIST Special Publication (SP) 800-160 Vol. 2 (Draft)). National Institute of Standards and Technology

United Kingdom (UK) Government (2010). A Strong Britain in an Age of Uncertainty: The National Security Strategy. Presented to Parliament by the Prime Minister by Command of Her Majesty. October 2010. ISBN: 9780101795326. Published by The Stationary Office (TSO), Norwich, UK. http://www.official-documents.gov.uk/

United States (U.S.) Air Force. (2015, September). *Air Force Future Operating Concept.*

U.S. Department of the Army. (2022, February). *United States Army Climate Strategy*, Office of the Assistant Secretary of the Army for Installations, Energy and Environment, Washington, DC.

U.S. Department of Defense (DoD). (2022). Agency Financial Report Fiscal Year 2022. Office of the Under Secretary of Defense. Washington, DC. http://comptroller.defense .gov/odcfo/AFR

U.S. Department of Homeland Security (DHS). (2022). FY22 Cybersecurity Sprints. https:// www.dhs.gov/cybersecurity-sprints (accessed December 29, 2022).

U.S. Department of the Navy. (2010). Department of the Navy's Energy Program for Security and Independence. Deputy Assistant Secretary of the Navy (DASN) Energy Office. http://www.navy.mil/secnav/ (accessed November 19, 2022).

U.S. Government. (2010). Telework Enhancement Act of 2010. Public Law 111–292 111th Congress December 9, 2010 https://www.telework.gov/guidance-legislation/telework -legislation/telework-enhancement-act/ (accessed November 28, 2022).

U.S. (2021). Guide to Telework and Remote Work in the Federal Government – Leveraging Telework and Remote Work in the Federal Government to Better Meet Our Human Capital Needs and Improve Mission Delivery. Office of Personnel Management. https://www.telework.gov/guidance-legislation/telework-guidance/telework-guide/

Vanek, F., L. Albright, and L. Angenent (2016). *Energy System Engineering: Evaluation and Implementation*, 3rd edition. McGraw-Hill Education, New York.

Williams, P. and J. Phillips (2011). *The Green Scorecard: Measuring the Return on Investment in Sustainability Initiatives*, Nicholas Brealey Publishing, Boston, MA/London.

World Economic Forum. (2022). *The Global Risks Report 2022*, 17th edition, ISBN: 978-2-940631-09-4. https://www.weforum.org/reports/global-risks-report-2022

World Resources Hub (WRI). (2022). *COP27 Resource Hub*. https://www.wri.org/un-climate -change-conference-resource-hub (accessed November 28, 2022).

7 Quantifying the Utility of Innovation

INTRODUCTION TO UTILITY MODELING

We, by default, believe that every innovation has utility to the same desirable extent. Unfortunately, this is not always the case. Every innovation has utility on differing scales and to differing extents. In fact, for social reasons, some innovations may not be desirable at all. For example, in warfare history, the invention of the Gatling Gun was hailed by some while abhorred by others. So, innovation can mean different things to different groups. For this reason, the technique of utility modeling is applicable for the assessment of innovation. The advantage of using utility modeling for innovation is that assessment metrics are imposed on the expected performance of innovation. How do we know if and when innovation has occurred and to what extent? It is easy to claim to be innovative or to proclaim commitment to innovation, but it is a different thing to be able to present a quantifiable proof of innovation.

INNOVATION INVESTMENT CHALLENGE

Innovation investment selection is an important aspect of investment planning. The right investment must be undertaken at the right time to satisfy the constraints of time and resources. A combination of criteria can be used to help in investment selection, including technical merit, management desire, schedule efficiency, cost–benefit ratio, resource availability, criticality of need, availability of sponsors, and user acceptance.

Many aspects of investment selection cannot be expressed in quantitative terms. For this reason, investment analysis and selection must be addressed by techniques that permit the incorporation of both quantitative and qualitative factors. Some techniques for investment analysis and selection are presented in the sections that follow. These techniques facilitate the coupling of quantitative and qualitative considerations in the investment decision process. Such techniques as net present value, profit ratio, and equity break-even point, which have been presented in the preceding chapters, are also useful for investment selection strategies.

UTILITY MODELS

The term "utility" refers to the rational behavior of a decision-maker faced with making a choice in an uncertain situation. The overall utility of an investment can be

DOI: 10.1201/9781003403548-7

measured in terms of both quantitative and qualitative factors. This section presents an approach to investment assessment based on utility models that have been developed within an extensive body of literature. The approach fits an empirical utility function to each factor that is to be included in a multi-attribute selection model. The specific utility values (weights) that are obtained from the utility functions are used as the basis for selecting an investment.

Utility theory is a branch of decision analysis that involves the building of mathematical models to describe the behavior of a decision-maker faced with making a choice among alternatives in the presence of risk. Several utility models are available in the management science literature. The utility of a composite set of outcomes of n decision factors is expressed in the general form:

$$U(x) = U(x_1, x_2, ... x_n),$$

where x_i = specific outcome of attribute X_i, $i = 1,2,...,n$, and $U(x)$ is the utility of the set of outcomes to the decision-maker. The basic assumption of utility theory is that people make decisions with the objective of maximizing those decisions' *expected utility.* Drawing on an example presented by Park and Sharp-Bette (1990), we may consider a decision-maker whose utility function with respect to investment selection is represented by the following expression:

$$u(x) = 1 - e^{-0.0001x}$$

where x represents a measure of the benefit derived from an investment. Benefit, in this sense, may be a combination of several factors (e.g., quality improvement, cost reduction, or productivity improvement) that can be represented in dollar terms. Suppose this decision-maker is faced with a choice between two investment alternatives, each of which has benefits as specified here:

Investment 1: Probabilistic levels of investment benefits.
  Benefits (x): −$10,000, $0, $10,000, $20,000, $30,000
  Probabilities (P(x)): 0.2, 0.2, 0.2, 0.2, 0.2

Investment 2: A definite benefit of $5,000.
  Assuming an initial benefit of zero and identical levels of required investment, the decision-maker must choose between the two investments. For Investment I, the expected utility is computed as follows:

$$E[u(x)] = \Sigma u(x)\{P(x)\}.$$

For Investment 1, using the utility function calculations and the probability values provided, we have the following series of calculations for the expected utility:

−\$10,000(−1.7183)(0.2) = −0.3437
\$0(0)(0.2) = 0
\$10,000(0.6321)(0.2) = 0.1264
\$20,000(0.8647)(0.2) = 0.1729
\$30,000(0.9502)(0.2) = 0.1900
Thus, $E[u(x)_1] = 0.1456$

For Investment 2, we have $u(x)_2 = u(\$5,000) = 0.3935$. Consequently, the investment providing the certain amount of \$5,000 is preferred to the riskier Investment 1, even though Investment 1 has a higher expected benefit of $\Sigma\, xP(x) = \$10,000$. If the expected utility of 0.1456 is set equal to the decision-maker's utility function, we obtain the following:

$$0.1456 = 1 - e^{-0.0001x*}$$

which yields $x* = \$1,574$, referred to as the *certainty equivalent (CE)* of Investment I $(CE_1 = 1,574)$. The CE of an alternative with variable outcomes is a *certain amount (CA)*, which a decision-maker will consider to be desirable to the same degree as the variable outcomes of the alternative. In general, if CA represents the certain amount of benefit that can be obtained from Investment II, then the criteria for making a choice between the two investments can be summarized as follows:

If $CA < \$1,574$, select Investment I
If $CA = \$1,574$, select either investment
If $CA > \$1,574$, select Investment II

The key in using utility theory for investment selection is choosing the proper utility model. The sections that follow describe two simple but widely used utility models: the *additive utility model* and the *multiplicative utility model*.

ADDITIVE UTILITY MODEL

The additive utility of a combination of outcomes of n factors $(X_1, X_2, \ldots, X_n)$ is expressed as follows:

$$U(x) = \sum_{i=1}^{n} U\left(x_i, \overline{x_i}^{-0}\right)$$

$$= \sum_{i=1}^{n} k_i U(x_i),$$

where

x_i = measured or observed outcome of attribute i
n = number of factors to be compared
x = combination of the outcomes of n factors

$U(x_i)$ = utility of the outcome for attribute i, x_i

$U(x)$ = combined utility of the set of outcomes, x

k_i = weight or scaling factor for attribute $i(0 < k_i < 1)$

X_i = variable notation for attribute i

x_i^0 = worst outcome of attribute i

x_i^* = best outcome of attribute i

$\overline{x}_i^{\,0}$ = set of worst outcomes for the complement of x_i

$U\left(x_i, \overline{x}_i^{\,0}\right)$ = utility of the outcome of attribute I and the set of worst outcomes

for the complement of attribute i

$$k_i = U\left(x_i^*, \overline{x}_i^{\,0}\right)$$

$$\sum_{i=1}^{n} k_i = 1.0 \text{(required for the additive model)}$$

Example 7.1: Let **A** be a collection of four investment attributes defined as **A** = {Profit, Flexibility, Quality, Productivity}. Now, define **X** = {Profit, Flexibility} as a subset of A. Then, $\overline{X}$ is the complement of **X**, defined as $\overline{X}$= {Quality, Productivity}. An example of the comparison of two investments under the additive utility model is summarized as:

$$U(x)_A = \sum_{i=1}^{n} k_i U_i(x_i) = .4(.95) + .2(.45) + .3(.35) + .1(.75) = 0.650$$

$$U(x)_B = \sum_{i=1}^{n} k_i U_i(x_i) = .4(.90) + .2(.98) + .3(.20) + .1(.10) = 0.626$$

since $U(x)_A > U(x)_B$, Investment A is selected.

MULTIPLICATIVE UTILITY MODEL

Under the multiplicative utility model, the utility of a combination of outcomes of n factors $(X_1, X_2, ..., X_n)$ is expressed as

$$U(x) = \frac{1}{C}\left[\prod_{i=1}^{n}\left(Ck_i U_i(x_i) + 1\right) - 1\right]$$

where C and k_i are scaling constants satisfying the following conditions:

$$\prod_{i=1}^{n}(1 + Ck_i) - C = 1.0$$

$$-1.0 < C < 0.0$$
$$0 < k_i < 1$$

The other variables are as defined previously for the additive model. Using the multiplicative model for an hypothetical case, involving Option A versus Option B, yields $U(x)_A = 0.682$ and $U(x)_B = 0.676$. Thus, Investment A is the better option.

FITTING A UTILITY FUNCTION

An approach presented in this section for multi-attribute investment selection is to fit an empirical utility function to each factor to be considered in the selection process. The specific utility values (weights) that are obtained from the utility functions may then be used in any of the standard investment justification methodologies. One way to develop empirical utility function for an investment attribute is to plot the "best" and "worst" outcomes expected from the attribute and then to fit a reasonable approximation of the utility function using concave, convex, linear, S-shaped, or any other logical functional form.

Alternately, if an appropriate probability density function can be assumed for the outcomes of the attribute, then the associated cumulative distribution function may yield a reasonable approximation of the utility values between 0 and 1 for corresponding outcomes of the attribute. In that case, the cumulative distribution function gives an estimate of the cumulative utility associated with increasing levels of attribute outcome. Simulation experiments, histogram plotting, and goodness-of-fit tests may be used to determine the most appropriate density function for the outcomes of a given attribute. For example, the following five attributes are used to illustrate how utility values may be developed for a set of investment attributes. The attributes are return on investment (ROI), productivity improvement, quality improvement, idle time reduction, and safety improvement.

Example 7.2: Suppose we have historical data on the ROI for investing in a particular investment. Assume that the recorded ROI values range from 0% to 40%. Thus, the worst outcome is 0% and the best outcome is 40%. A frequency distribution of the observed ROI values is developed, and an appropriate probability density function (*pdf*) is fitted to the data. For our example, suppose the ROI is found to be exceptionally distributed with a mean of 12.1%. That is:

$$f(x) = \begin{cases} \dfrac{1}{\beta} e^{-x/\beta}, & if \quad x \geq 0 \\ 0, & otherwise \end{cases}$$

$$F(x) = \begin{cases} 1 - e^{-x/\beta}, & if \quad x \geq 0 \\ 0, & otherwise \end{cases}$$

$$\approx U(x),$$

where $\beta = 12.1$. $F(x)$ approximates $U(x)$. The utility of any observed ROI within the applicable range may be read directly from the cumulative distribution function. For the productivity improvement attribute, suppose it is found (based on historical data analysis) that the level of improvement is normally distributed with a mean of 10% and a standard deviation of 5%. That is,

$$f(x) = \frac{1}{\sqrt{2\pi}\sigma} e^{-\frac{1}{2}\left(\frac{x-\mu}{\sigma}\right)^2}, \quad -\infty < x < \infty$$

where $\pi = 10$ and $\sigma = 5$. Since the normal distribution does not have a closed-form expression for $F(x)$, $U(x)$ is estimated by plotting representative values based on the standard normal table. The utility of productivity improvement may also be evaluated on the basis of cost reduction. Suppose quality improvement is subjectively assumed to follow a beta distribution with shape parameters $\alpha = 1.0$ and $\beta = 2.9$. That is,

$$f(x) = \frac{\Gamma(\alpha+\beta)}{\Gamma(\alpha)\Gamma(\beta)} \cdot \frac{1}{(b-a)^{\alpha+\beta-1}} \cdot (x-a)^{\alpha-1}(b-x)^{\beta-1},$$

$$\text{for } a \le x \le b \text{ and } \alpha > 0, \ \beta > 0.$$

where

    a = lower limit for the distribution
    b = upper limit for the distribution
    α, β are the shape parameters for the distribution

As with the normal distribution, there is no closed-form expression for $F(x)$ for the beta distribution. However, if either of the shape parameters is a positive integer, then a binomial expansion can be used to obtain $F(x)$. Based on work analysis observations, suppose idle time reduction is found to be best described by a log normal distribution with a mean of 10% and standard deviation of 5%. This is represented as follows:

$$f(x) = \frac{1}{x\sqrt{2\pi\sigma^2}} e^{\left[\frac{-(\ln x-\mu)^2}{2\sigma^2}\right]}, \quad x > 0$$

For the example, suppose safety improvement is assumed to have a previously known utility function, defined as follows:

$$U_p(x) = 30 - \sqrt{400 - x^2}$$

where x represents percentage improvement in safety. For the expression, the unscaled utility values range from 10 (for 0% improvement) to 30 (for 20% improvement). To express any particular outcome of an attribute i, x_i, on a scale of 0.0 to 1.0, it is expressed as a proportion of the range of best to worst outcomes as shown here:

$$X = \frac{x_i - x_i^0}{x_i^* - x_i^0}$$

where X = outcome expressed on a scale of 0.0 to 1.0

x_i = measured or observed raw outcome of attribute i

x_i^0 = worst raw outcome of attribute i

x_i^* = best raw outcome of attribute i.

The utility of the outcome may then be represented as $U(X)$ and read off the empirical utility curve. Using this approach, the utility function for safety improvement is scaled from 0.0 to 1.0. The respective utility values for a system's attributes may be viewed as relative weights for comparing investment alternatives. The utility obtained from the modeled functions can be used in the additive and multiplicative utility models discussed earlier. Using the additive utility model, the *composite utility (CU)* of the investment, based on the five attributes, is given by:

$$U(X) = \sum_{i=1}^{n} k_i U_i(x_i)$$

$$= .30(.61) + .20(.49) + .25(.93) + .15(.86) + .10(.40) = 0.6825.$$

This CU value may then be compared with the utilities of other investments. On the other hand, a single investment may be evaluated independently on the basis of some minimum acceptable level of utility (MALU) desired by the decision-maker. The criteria for evaluating an investment based on MALU may be expressed by the following rule:

Investment j is acceptable if its composite utility, $U(X)_j$, is greater than MALU
Investment j is not acceptable if its composite utility, $U(X)_j$, is less than MALU

The utility of an investment may be evaluated on the basis of its economic, operational, or strategic importance to an organization. Utility functions can be incorporated into existing justification methodologies. For example, in the analytic hierarchy process, utility functions can be used to generate values that are, in turn, used to evaluate the relative preference levels of attributes and alternatives. Utility functions can be used to derive component weights when the overall effectiveness of investments is being compared. Utility functions can generate descriptive levels of investment performance as well as indicating the limits of innovation investment effectiveness.

POLAR PLOTS FOR INNOVATION ASSESSMENT

Polar plots provide a means of visually comparing investment alternatives (Badiru, 1991). In a conventional polar plot, the vectors drawn from the center of the circle are on individual scales based on the outcome ranges for each attribute. For example, the vector for NPV (Net Present Value) is on a scale of $0 to $500,000, while the scale for quality is from 0 to 10. It should be noted that the overall priority weights for the alternatives are not proportional to the areas of their respective polyhedrons. A modification of the basic polar plot is presented in this section. The modification involves a procedure that normalizes the areas of the polyhedrons with respect to the total area of the base circle. With this modification, the normalized areas of the polyhedrons are proportional to the respective priority weights of the alternatives, so the alternatives can be ranked on the basis of the areas of the polyhedrons. The steps involved in the modified approach are presented as follows:

1. Let n be the number of attributes involved in the comparison of alternatives, such that $n \geq 4$. Number the attributes in a preferred order $(1,2,3, \ldots, n)$.
2. If the attributes are considered to be equally important (i.e., equally weighted), compute the sector angle associated with each attribute as

$$\theta = \frac{360^\circ}{n}$$

3. Draw a circle with a large enough radius. A radius of 2 inches is usually adequate.
4. Convert the outcome range for each attribute to a standardized scale of 0 to 10 using appropriate transformation relationships.
5. For Attribute 1, draw a vertical vector up from the center of the circle to the edge of the circle.
6. Measure θ clockwise and draw a vector for Attribute 2. Repeat this step for all attributes in the numbered order.
7. For each alternative, mark its standardized relative outcome with respect to each attribute along the attribute's vector. If a 2-inch radius is used for the base circle, then we have the following linear transformation relationship:
   0.0 inches = rating score of 0.0
   2.0 inches = rating score of 10.0
8. Connect the points marked for each alternative to form a polyhedron. Repeat this step for all alternatives.

9. Compute the area of the base circle as follows:

$$\Omega = \pi r^2$$

$$= 4\pi \text{ squared inches}$$

$$= 100\pi \text{ squared rating units}$$

10. Compute the area of the polyhedron corresponding to each alternative. This can be done by partitioning each polyhedron into a set of triangles and then calculating the areas of the triangles. To calculate the area of each triangle, note that we will know the lengths of two sides of the triangle and the angle subtended by the two sides. With these three known values, the area of each triangle can be calculated through basic trigonometric formulas.

For example, the area of each polyhedron may be represented as λ_1 ($I = 1, 2, ..., m$), where m is the number of alternatives. The area of each triangle in the polyhedron for a given alternative is then calculated as

$$\Delta_t = \frac{1}{2}(L_j)(L_{j+1})(\sin\theta)$$

where

L_j = standardized rating with respect to attribute j
L_{j+1} = standardized rating with respect to attribute $j+1$
L_j and L_{j+1} are the two sides that subtend θ

Since $n \geq 4$, θ will be between 0 and 90 degrees, and $\sin(\theta)$ will be strictly increasing over that interval.

The area of the polyhedron for alternative i is then calculated as

$$\lambda_i = \sum_{t(i)=1}^{n} \Delta_{t(i)}$$

Note that θ is constant for a given number of attributes, and the area of the polyhedron will be a function of the adjacent ratings (L_j and L_{j+1}) only.

11. Compute the standardized area corresponding to each alternative as

$$w_i = \frac{\lambda_i}{\Omega}(100\%)$$

12. Rank the alternatives in decreasing order of λ_i. Select the highest-ranked alternative as the preferred alternative.

As an extension to the modification presented here, the sector angle may be a variable indicating relative attribute weights, while the radius represents the evaluation rating of the alternatives with respect to the weighted attribute. That is, if the attributes are not equally weighted, the sector angles will not all be equal. In that case, the sector angle for each attribute is computed as

$$\theta_j = p_j\left(360°\right)$$

where

p_j = relative numeric weight of each of n attributes

$$\sum_{j=1}^{n} p_j = 1.0$$

It should be noted that the weighted areas for the alternatives are sensitive to the order in which the attributes are drawn in the polar plot. Thus, a preferred order of the attributes must be defined prior to starting the analysis. The preferred order may be based on the desired sequence in which alternatives must satisfy management goals. For example, it may be desirable to attend to product quality issues before addressing throughput issues. The surface area of the base circle may be interpreted as a measure of the global organizational goal with respect to such performance indicators as available capital, market share, capacity utilization, and so on. Thus, the weighted area of the polyhedron associated with an alternative may be viewed as the degree to which that alternative satisfies organizational goals.

Some of the attributes involved in a selection problem might constitute a combination of quantitative and/or qualitative factors or a combination of objective and/or subjective considerations. The prioritization of the factors and considerations is typically based on the experience, intuition, and subjective preferences of the decision-maker. Goal programming is another technique that can be used to evaluate multiple objectives or criteria in decision problems.

INNOVATION BENCHMARKING

The techniques presented in the preceding sections can be used for benchmarking investments. For example, to develop a baseline schedule, evidence of successful practices from other investments may be needed. Metrics based on an organization's most critical investment implementation issues should be developed. *Benchmarking* is a process whereby target performance standards are established based on the best examples available. The objective is to equal or surpass the best example. In its simplest term, benchmarking means learning from and emulating a superior example. The premise of benchmarking is that if an organization replicates the best-quality examples, it will become one of the best in the industry. A major approach of benchmarking is to identify performance gaps between investments. Benchmarking requires that an attempt be made to close the gap by improving the performance of the subject investment.

Benchmarking requires frequent comparison with the target investment. Updates must be obtained from investments already benchmarked, and new investments to be benchmarked must be selected on a periodic basis. Measurement, analysis, feedback, and modification should be incorporated into the performance improvement program.

The feedback information of assessment metrics is necessary to determine what control actions should be taken at the next innovation improvement phase. The primary responsibility of an economic analyst is to ensure proper forward and backward flow of information concerning the performance of an investment on the basis of the benchmarked inputs.

USING LEARNING CURVES TO QUANTIFY SUSTAINABILITY OF INNOVATION

Learning curves, also known as manufacturing improvement curves, are used extensively in business, science, technology, engineering, and industry to predict system performance over time. For the purpose of this book, we adapt the conventional manufacturing improvement curve to a technology improvement curve with alignment to the needs of Industry 4.0 and Systems 4.0. Although most of the early development and applications were in the area of production engineering, contemporary applications can be found in all areas of application. Manufacturing, in particular, offers a fertile area for the application of learning curves. This chapter applies the concept of half-life of learning curves to manufacturing technology project management. This is useful for predictive measures of manufacturing system performance. Half-life is the amount of time it takes for a quantity to diminish to half of its original size through natural processes. The approach of half-life computation provides an additional decision tool for researchers and practitioners in manufacturing. Derivation of the half-life equations of learning curves can reveal more about the properties of the various curves with respect to the unique life cycle property of manufacturing enterprises.

Manufacturing technology has several unique characteristics that make its management challenging. Some of these characteristics include frequent life cycle changes and uncertainty in the operating environment. Learning curve analysis offers a viable approach for evaluating manufacturing systems where human learning and forgetting are involved. With an effective learning curve evaluation, an assessment can be made of how a manufacturing technology project meets organizational objectives and maximizes its benefits to the organization. It is essential to be able to predict locations of dips in the improvement curve so that an accurate assessment of the overall technology performance can be done. The half-life theory of learning curves introduced by Badiru and Ijaduola (2009) offers one good technique for an accurate assessment.

The fact is that the fast pace of technology affects learning curves, and the frequent changes of technology degrade learning curves. Thus, specialized analytical assessment of learning curves is needed for manufacturing technology project management. The degradation of learning curves is often depicted analytically by incorporating forgetting components into conventional learning curves, as has been shown in the literature over the past few decades (Badiru, 1994, 1995a; Jaber and Sikstrom, 2004; Jaber et al., 2003; Jaber and Bonney, 1996; Jaber and Bonney, 2007; Nembhard and Osothsilp, 2001; Nembhard and Uzumeri, 2000; Sule, 1978; Globerson et al., 1998).

HALF-LIFE THEORY OF LEARNING CURVES

As another approach to capturing the essence of forgetting in learning curves, Badiru and Ijaduola (2009) introduced the half-life theory of learning curves. Traditionally, the standard time has been used as an indication of when learning should cease or when resources need to be transferred to another job. It is possible that half-life theory can supplement standard time analysis. The half-life approach will encourage researchers and practitioners to reexamine conventional applications of existing learning curve models. Organizations invest in people, work process, and technology for the purpose of achieving performance improvement. The systems nature of such investment strategy requires that the investment be strategically planned over multiple years. Thus, changes in learning curve profiles over those years become crucial. Forgetting analysis and half-life computations can provide additional insights into learning curve changes. Through the application of robust learning curve analysis, system enhancement can be achieved in terms of cost, time, and performance with respect to strategic investment of funds and other organizational assets in people, process, and technology. The predictive capability of learning curves is helpful in planning for integrated system performance improvement.

Formal analysis of learning curves first emerged in the mid-1930s in connection with the analysis of the production of airplanes (Wright, 1936). Learning refers to the improved operational efficiency and cost reduction obtained from repetition of a task. This has a direct impact for training purposes and the design of work. Workers learn and improve by repeating operations. But, they also regress due to the impact of forgetting, prolonged breaks, work interruption, and natural degradation of performance. Half-life computations can provide a better understanding of actual performance levels over time. Half-life is the amount of time it takes for a quantity to diminish to half of its original size through natural processes. Duality is of natural interest in many real-world processes. We often speak of "twice as much" and "half as much" as benchmarks for process analysis. In economic and financial principles, the "rule of 72" refers to the length of time required for an investment to double in value. These common "double" or "half" concepts provide the motivation for half-life analysis.

The usual application of half-life is in natural sciences. For example, in physics, the half-life is a measure of the stability of a radioactive substance. In practical terms, the half-life attribute of a substance is the time it takes for one-half of the atoms in an initial magnitude to disintegrate. The longer the half-life of a substance, the more stable it is. This provides a good analogy for modeling learning curves with the recognition of increasing performance or decreasing cost with respect to the passage of time. The approach provides another perspective to the large body of literature on learning curves. Badiru and Ijaduola (2009) present the following formal definitions:

> *For learning curves: Half-life* is the production level required to reduce cumulative average cost per unit to half of its original size.
> *For forgetting curve: Half-life* is the amount of time it takes for performance to decline to half its original magnitude.

HUMAN–TECHNOLOGY PERFORMANCE DEGRADATION

Although there is an extensive collection of classical studies of *improvement* due to learning curves, only very limited attention has been paid to performance *degradation* due to the impact of forgetting. Some of the classical works on process improvement due to learning include Belkaoui (1976), Camm et al. (1987), Liao (1979), Mazur and Hastie (1978), McIntyre (1977), Nanda (1979), Pegels (1976), Richardson (1978), Smith (1989), Smunt (1986), Sule (1978), Womer (1979, 1981, 1984), Womer and Gulledge (1983), and Yelle (1976, 1979, 1983). It is only in recent years that the recognition of "forgetting" curves has begun to emerge, as can be seen in more recent literature (Badiru, 1995a; Jaber and Sikstrom, 2004; Jaber et al., 2003; Jaber and Bonney, 2003, 2007; Jaber and Guiffrida, 2008). The new and emerging research on the forgetting components of learning curves provides the motivation for studying half-life properties of learning curves. Performance decay can occur due to several factors, including lack of training, reduced retention of skills, lapses in performance, extended breaks in practice, and natural forgetting. The conventional learning curve equation introduced by Wright (1936) has a drawback whereby the cost/time per unit approaches zero as the cumulative output approaches infinity. That is:

$$\lim_{x \to \infty} C(x) = \lim_{x \to \infty} C_1 x^{-b} \to 0$$

Researchers who initially embraced Wright's learning curve (WLC) assumed a lower bound for the equation, such that WLC could be represented as:

$$C(x) = \begin{cases} C_1 x^{-b}, \text{if } x < x_s \\ C_s \quad, \text{otherwise} \end{cases}$$

where x_s is the number of units required to reach standard cost C_s. A half-life analysis can reveal more information about the properties of WLC, particularly when we consider the operating range of $x_0 < x_s$.

HALF-LIFE DERIVATIONS

Learning curves present the relationship between cost (or time) and level of activity on the basis of the effect of learning. An early study by Wright (1936) disclosed the "80 percent learning" effect, which indicates that a given operation is subject to a 20% productivity improvement each time the activity level or production volume *doubles*. The proposed half-life approach is the antithesis of the double-level milestone. Learning curve can serve as a predictive tool for obtaining time estimates for tasks that are repeated within a project life cycle. A new learning curve does not necessarily commence each time a new operation is started, since workers can sometimes transfer previous skills to new operations. The point at which the learning curve begins to flatten depends on the degree of similarity of the new operation to previously performed operations. Typical learning rates that have been encountered in practice range from 70% to 95%. Several alternate models of learning curves have been presented in the literature, including *log-linear model, S-curve model, Stanford-B model, DeJong's learning formula, Levy's adaptation function, Glover's learning formula, Pegels' exponential function, Knecht's upturn model*, and *Yelle's product model*. The basic log-linear model is the most popular learning curve model. It expresses a dependent variable (e.g., production cost) in terms of some independent variable (e.g., cumulative production). The model states that the improvement in productivity is constant (i.e., it has a constant slope) as output increases. That is:

$$C(x) = C_1 x^{-b}$$

where:

$C(x)$ = cumulative average cost of producing x units
C_1 = cost of the first unit
x = cumulative production unit
b = learning curve exponent

The expression for $C(x)$ is practical only for $x > 0$. This makes sense, because learning effect cannot realistically kick in until at least one unit ($x \geq 1$) has been produced. For the standard log-linear model, the expression for the learning rate, p, is derived by considering two production levels where one level is double the other. The performance curve, $P(x)$, can be defined as the reciprocal of the average cost curve, $C(x)$. Thus, we have:

$$P(x) = \frac{1}{C(x)}$$

which will have an increasing profile compared with the asymptotically declining cost curve. In terms of practical application, learning to drive is one example, where maximum performance can be achieved in a relatively short time compared with the half-life of performance. That is, learning is steep, but the performance curve is relatively flat after steady state is achieved. The application of half-life analysis to learning curves can help address questions such as these:

- How fast and how far can system performance be improved?
- What are the limitations to system performance improvement?
- How resilient is a system to shocks and interruptions to its operation?
- Are the performance goals that are set for the system achievable?

HALF-LIFE OF THE LOG-LINEAR MODEL

The half-life of the log-linear model is computed as follows. Let:

C_0 = Initial performance level
$C_{1/2}$ = Performance level at half-life

$$C_0 = C_1 x_0^{-b} \quad \text{and} \quad C_{1/2} = C_1 x_{1/2}^{-b}$$

But, $C_{1/2} = \frac{1}{2} C_0$

Therefore, $C_1 x_{1/2}^{-b} = \frac{1}{2} C_1 x_0^{-b}$, which leads to $x_{1/2}^{-b} = \frac{1}{2} x_0^{-b}$, which, by taking the $(-1/b)$th exponent of both sides, simplifies to yield the following expression as the general expression for the standard log-linear learning curve model:

$$x_{1/2} = \left(\frac{1}{2} \right)^{-\frac{1}{b}} x_0, \quad x_0 \geq 1$$

where $x_{1/2}$ is the half-life and x_0 is the initial point of operation. We refer to $x_{1/2}$ as the **first-order half-life**. The **second-order half-life** is computed as the time corresponding to half of the preceding half. That is:

$$C_1 x_{1/2(2)}^{-b} = \frac{1}{4} C_1 x_0^{-b}$$

which simplifies to yield:

$$x_{1/2(2)} = (1/2)^{-2/b} x_0$$

Similarly, the **third-order half-life** is derived to obtain:

$$x_{1/2(3)} = (1/2)^{-3/b} x_0$$

In general, the **kth-order half-life** for the log-linear model is represented as:

$$x_{1/2(k)} = (1/2)^{-k/b} x_0$$

HALF-LIFE COMPUTATIONAL EXAMPLES

This section uses examples of log-linear learning curves with $b = 0.75$ and $b = 0.3032$, respectively, to illustrate the characteristics of learning that can dictate the half-life behavior of the overall learning process. Knowing the point where the half-life of each curve occurs can be very useful in assessing learning retention for the purpose of designing training programs or designing work. For $C(x) = 250x^{-0.75}$, the first-order half-life is computed as:

$$x_{1/2} = (1/2)^{-1/0.75} x_0, \quad x_0 \geq 1$$

If this expression is evaluated for $x_0 = 2$, the first-order half-life yields $x_{1/2} = 5.0397$, which indicates a fast drop in the value of $C(x)$. $C(2) = 148.6509$, corresponding to a half-life of 5.0397. Note that $C(5.0397) = 74.7674$, which is about half of 148.6509. The conclusion from this analysis is that if we are operating at the point $x = 2$, we can expect the curve to reach its half-life decline point at $x = 5$. For $C(x) = 240.03x^{-0.3032}$, the first-order half-life is computed as:

$$x_{1/2} = (1/2)^{-1/0.3032} x_0, \quad x_0 \geq 1$$

If we evaluate this function for $x_0 = 2$, the first-order half-life is $x_{1/2} = 19.6731$. Several models and variations of learning curves are used in practice. Models are developed through one of the following approaches:

1. Conceptual models
2. Theoretical models
3. Observational models
4. Experimental models
5. Empirical models

The S-Curve Model: The S-Curve (Towill and Kaloo, 1978) is based on the assumption of a gradual start-up. The function has the shape of the cumulative normal distribution function for the start-up curve and the shape of an operating characteristics function for the learning curve. The gradual start-up is based on the fact that the early stages of production are typically in a transient state with changes in tooling, methods, materials, design, and even changes in the work force. The basic form of the S-Curve function is:

$$C(x) = C_1 + M(x+B)^{-b}$$

$$MC(x) = C_1 \left[M + (1-M)(x+B)^{-b} \right]$$

where:

$C(x)$ = learning curve expression
b = learning curve exponent
$M(x)$ = marginal cost expression
C_1 = cost of first unit
M = incompressibility factor (a constant)
B = equivalent experience units (a constant)

Assumptions about at least three out of the four parameters (M, B, C_1, and b) are needed to solve for the fourth one. Using the $C(x)$ expression and derivation procedure outlined earlier for the log-linear model, the half-life equation for the S-Curve learning model is derived to be:

$$x_{1/2} = (1/2)^{-1/b} \left[\frac{M(x_0+B)^{-b} - C_1}{M} \right]^{-1/b} - B$$

where:

$x_{1/2}$ = half-life expression for the S-Curve learning model
x_0 = initial point of evaluation of performance on the learning curve

In terms of practical application of the S-Curve, consider when a worker begins learning a new task. The individual is slow initially at the tail end of the S-Curve. But, the rate of learning increases as time goes on, with additional repetitions. This helps the worker to climb the steep-slope segment of the S-Curve very rapidly. At the top of the slope, the worker is classified as being proficient with the learned task. From then on, even if the worker puts much effort into improving upon the task, the resultant learning will not be proportional to the effort expended. The top end of the S-curve is often called the slope of *diminishing returns*. At the top of the S-Curve, workers succumb to the effects of *forgetting* and other performance-impeding factors. As the work environment continues to change, a worker's level of skill and expertise can become obsolete. This is an excellent reason for the application of half-life computations.

The Stanford-B Model: The Stanford-B model is represented as:

$$UC(x) = C_1 (x + B)^{-b}$$

where:

$UC(x)$ = direct cost of producing the xth unit
b = learning curve exponent
C_1 = cost of the first unit when $B = 0$
B = slope of the asymptote for the curve
B = constant$(1 < B < 10)$. This is equivalent to units of previous experience at the start of the process, which represents the number of units produced prior to first unit acceptance. It is noted that when $B = 0$, the Stanford-B model reduces to the conventional log-linear model. The general expression for the half-life of the Stanford-B model is derived to be:

$$x_{1/2} = (1/2)^{-1/b}(x_0 + B) - B$$

where:

$x_{1/2}$ = half-life expression for the Stanford-B learning model
x_0 = initial point of evaluation of performance on the learning curve

Badiru's Multi-Factor Model: Badiru (1994) presents applications of learning and forgetting curves to productivity and performance analysis. One example presented used production data to develop a predictive model of production throughput. Two data replicates are used for each of ten selected combinations of cost and time values. Observations were recorded for the number of units representing double production levels. The resulting model has the following functional form:

$$C(x) = 298.88x_1^{-0.31}x_2^{-0.13}$$

where:

$C(x)$ = cumulative production volume
x_1 = cumulative units of Factor 1
x_2 = cumulative units of Factor 2
b_1 = first learning curve exponent = -0.31
b_2 = second learning curve exponent = -0.13

A general form of the modeled multi-factor learning curve model is:

$$C(x) = C_1 x_1^{-b_1} x_2^{-b_2}$$

and the half-life expression for the multi-factor learning curve was derived to be:

$$x_{1(1/2)} = (1/2)^{-1/b_1} \left[\frac{x_{1(0)} x_{2(0)}^{b_2/b_1}}{x_{2(1/2)}^{b_2/b_1}} \right]^{-1/b_1}$$

$$x_{2(1/2)} = (1/2)^{-1/b_2} \left[\frac{x_{2(0)} x_{1(0)}^{b_1/b_2}}{x_{1(1/2)}^{b_2/b_1}} \right]^{-1/b_2}$$

where:

$x_{i(1/2)}$ = half-life component due to Factor i (i=1, 2)
$x_{i(0)}$ = initial point of Factor i (i=1, 2) along the multi-factor learning curve

Knowledge of the value of one factor is needed to evaluate the other factor. Just as in the case of single-factor models, the half-life analysis of the multi-factor model can be used to predict when the performance metric will reach half of a starting value.

DeJong's Learning Formula: DeJong's learning formula is a power function that incorporates parameters for the proportion of manual activity in a task. When operations are controlled by manual tasks, the time will be compressible as successive units are completed. If, by contrast, machine cycle times control operations, then the time will be less compressible as the number of units increases. DeJong's formula introduces an incompressible factor, M, into the log-linear model to account for the man–machine ratio. The model is expressed as:

$$C(x) = C_1 + M x^{-b}$$

$$MC(x) = C_1 \left[M + (1 - M) x^{-b} \right]$$

where:

$C(x)$ = learning curve expression
$M(x)$ = marginal cost expression
b = learning curve exponent
C_1 = cost of first unit
M = incompressibility factor (a constant)

When $M = 0$, the model reduces to the log-linear model, which implies a completely manual operation. In completely machine-dominated operations, $M = 1$. In that case, the unit cost reduces to a constant equal to C_1, which suggests that no learning-based cost improvement is possible in machine-controlled operations. This represents a condition of high incompressibility. This profile suggests impracticality at higher values of production. Learning is very steep, and average cumulative production cost drops rapidly. The horizontal asymptote for the profile is below the lower bound on the average cost axis, suggesting an infeasible operating region as production volume becomes high. This analysis agrees with the fact that no significant published data is available on whether or not DeJong's learning formula has been successfully used to account for the degree of automation in any given operation. Using the expression $MC(x)$, the marginal cost half-life of the DeJong's learning model is derived to be:

$$x_{1/2} = (1/2)^{-1/b} \left[\frac{(1-M)x_0^{-b} - M}{2(1-M)} \right]^{-1/b}$$

where:

$x_{1/2}$ = half-life expression for the DeJong's learning curve marginal cost model
x_0 = initial point of evaluation of performance on the marginal cost curve

If the $C(x)$ model is used to derive the half-life, then we obtain the following derivation:

$$x_{1/2} = (1/2)^{-1/b} \left[\frac{Mx_0^{-b} - C_1}{M} \right]^{-1/b}$$

where:

$x_{1/2}$ = half-life expression for the DeJong's learning curve model
x_0 = initial point of evaluation of performance on the DeJong's learning curve

Levy's Technology Adaptation Function: Recognizing that the log-linear model does not account for leveling off of production rate and the factors that may influence learning, Levy (1965) presented the following learning cost function:

$$MC(x) = \left[\frac{1}{\beta} - \left(\frac{1}{\beta} - \frac{x^{-b}}{C_1} \right) k^{-kx} \right]^{-1}$$

where:

β = production index for the first unit
k = constant used to flatten the learning curve for large values of x

The flattening constant, k, forces the curve to reach a plateau instead of continuing to decrease or turning in the upward direction. The half-life expression for Levy's learning model is a complex nonlinear expression derived as follows:

$$(1/\beta - x_{1/2}^{-b}/C_1)k^{-kx_{1/2}} = 1/\beta - 2[1/\beta - (1/\beta - x_0^{-b}/C_1)k^{-kx_0}]$$

where:

$x_{1/2}$ = half-life expression for the Levy's learning curve model
x_0 = initial point of evaluation of performance on the Levy's learning curve

Knowledge of some of the parameters of the model is needed to solve for the half-life as a closed-form expression.

Glover's Learning Model: Glover's learning formula (Glover, 1966) is a learning curve model that incorporates a work commencement factor. The model is based on a bottom-up approach, which uses individual worker learning results as the basis for plant-wide learning curve standards. The functional form of the model is expressed as:

$$\sum_{i=1}^{n} y_i + a = C_1 \left(\sum_{i=1}^{n} x_i \right)^m$$

where:

y_i = elapsed time or cumulative quantity
x_i = cumulative quantity or elapsed time
a = commencement factor
n = index of the curve (usually $1 + b$)
m = model parameter

This is a complex expression for which half-life expression is not easily computable. We defer the half-life analysis of Levy's learning curve model for further research by interested readers.

Pegels' Exponential Function: Pegels (1976) presented an alternate algebraic function for the learning curve. His model, a form of an exponential function of marginal cost, is represented as:

$$MC(x) = \alpha a^{x-1} + \beta$$

where α, β, and a are parameters based on empirical data analysis. The total cost of producing x units is derived from the marginal cost as follows:

$$TC(x) = \int \left(\alpha a^{x-1} + \beta \right) dx = \frac{\alpha a^{x-1}}{\ln(a)} + \beta x + c$$

where c is a constant to be derived after the other parameters are found. The constant can be found by letting the marginal cost, total cost, and average cost of the first unit be all equal. That is, $MC_1 = TC_1 = AC_1$, which yields:

$$c = \alpha - \frac{\alpha}{\ln(a)}$$

The model assumes that the marginal cost of the first unit is known. Thus,

$$MC_1 = \alpha + \beta = y_0$$

Mathematical expression for the total labor cost in Pegels' start-up curves is expressed as:

$$TC(x) = \frac{a}{1-b} x^{1-b}$$

where:

x = cumulative number of units produced
a, b = empirically determined parameters

The expressions for marginal cost, average cost, and unit cost can be derived as shown earlier for other models. Using the total cost expression, $TC(x)$, we derive the expression for the half-life of Pegels' learning curve model to be as shown here:

$$x_{1/2} = (\tfrac{1}{2})^{-1/(1-b)} x_0$$

Knecht's Upturn Model: Knecht (1974) presents a modification to the functional form of the learning curve to analytically express the observed divergence of actual costs from those predicted by learning curve theory when units produced exceed 200. This permits the consideration of non-constant slopes for the learning curve model. If UC_x is defined as the unit cost of the xth unit, then it approaches 0 asymptotically as x increases. To avoid a zero limit unit cost, the basic functional form is modified. In the continuous case, the formula for cumulative average costs is derived as:

$$C(x) = \int_0^x C_1 z^b dz = \frac{C_1 x^{b+1}}{(1+b)}$$

This cumulative cost also approaches zero as x goes to infinity. Knecht alters the expression for the cumulative curve to allow for an upturn in the learning curve at large cumulative production levels. He suggested the following functional form:

$$C(x) = C_1 x^{-b} e^{cx}$$

where c is a second constant. Differentiating the modified cumulative average cost expression gives the unit cost of the xth unit as shown in the following. This formulation shows the cumulative average cost plot of Knecht's upturn function for values of $C_1 = 250$, $b = 0.25$, and $c = 0.25$.

$$UC(x) = \frac{d}{dx}\left[C_1 x^{-b} e^{cx} \right] = C_1 x^{-b} e^{cx}\left(c + \frac{-b}{x} \right)$$

The half-life expression for Knecht's learning model turns out to be a nonlinear complex function, as shown here:

$$x_{1/2}e^{-cx_{1/2}/b} = (\tfrac{1}{2})^{-1/b}e^{-cx_0/b}x_0$$

where:

$x_{1/2}$ = half-life expression for the Knecht's learning curve model
x_0 = initial point of evaluation of performance on the Knecht's learning curve

Given that x_0 is known, iterative, interpolation, or numerical methods may be needed to solve for the half-life value.

Yelle's Combined Technology Learning Curve: Yelle (1979) proposed a learning curve model for products by aggregating and extrapolating the individual learning curve of the operations making up a product on a log-linear plot. The model is expressed as shown here:

$$C(x) = k_1 x_1^{-b_1} + k_2 x_2^{-b_2} + \cdots + k_n x_n^{-b_n}$$

Where:

$C(x)$ = cost of producing the xth unit of the product
n = number of operations making up the product
$k_i x_i^{-b_i}$ = learning curve for the ith operation

Aggregated Learning Curves: In comparing the models discussed in the preceding sections, the deficiency of Knecht's model is that a product-specific learning curve seems to be a more reasonable model than an integrated product curve. For example, an aggregated learning curve with 96.6% learning rate obtained from individual learning curves with the respective learning rates of 80%, 70%, 85%, 80%, and 85% does not appear to represent reality. If this type of composite improvement is possible, then one can always improve the learning rate for any operation by decomposing it into smaller integrated operations. The additive and multiplicative approaches of reliability functions support the conclusion of impracticality of Knecht's integrated model. Jaber and Guiffrida (2004)presented an aggregated form of the WLC where some of the items produced are defective and require reworking. The quality learning curve that they provide is of the form:

$$t(x) = y_1 x^{-b} + 2r_1 \left(\frac{p}{2}\right)^{1-\varepsilon} x^{1-2\varepsilon}$$

where y_1 is the time to produce the first unit, r_1 is the time to rework the first defective unit, p is the probability of the process to go out-of-control ($p \ll 1$), and b is the learning exponent of the reworks learning curve. The variable $t(x)$ has three behavioral patterns, for $0 < b < \frac{1}{2}$ (Case I), $b = \frac{1}{2}$ (Case II), and $\frac{1}{2} < b < 1$ (Case III). Assuming no production error, we computed the half-life for $t(x)$ for case I as:

Case I: $x_{1/2} = \left(\dfrac{1}{2}\right)^{-\frac{1}{b}} x$ and $x_{1/2} = \left(\dfrac{1}{2}\right)^{-\frac{1}{1-2\varepsilon}} x$

Case II: $t(x) = y_1 x^{-b} + 2r_1\left(\dfrac{\rho}{2}\right)^{1-\varepsilon} x^{1-2\varepsilon}$ reduces to $t(x) = y_1 x^{-b} + t(x) = y_1 x^{-b} + 2r_1\sqrt{\dfrac{\rho}{2}}$,

where $2r_1\sqrt{\dfrac{\rho}{2}}$ is the lower bound, or the plateau of the learning curve.

Case III: The behavior of $t(x)$ follows that of the WLC: monotonically decreasing as cumulative output increases. It is noted that Jaber and Guiffrida (2008) assumed that the percentage defective reduces as the number of interruptions to restore the process increases. They found that $t(x)$ could converge to the WLC as the learning curve exponent becomes insignificant.

HALF-LIFE OF DECLINE CURVES

Over the years, the decline curve technique has been extensively used by the oil and gas industry to evaluate future oil and gas predictions. These predictions are used as the basis for economic analysis to support development, property sale or purchase, industrial loan provisions, and also to determine if a secondary recovery project should be carried out. It is expected that the profile of a hyperbolic decline curve can be adapted for application to learning curve analysis. The graphical solution of the hyperbolic equation is through the use of a log-log paper, which sometimes provides a straight line that can be extrapolated for a useful length of time to predict future production levels. This technique, however, sometimes failed to produce the straight line needed for extrapolation for some production scenarios. Furthermore, the graphical method usually involves some manipulation of data, such as shifting, correcting, and/or adjusting scales, which eventually introduces bias into the actual data. In order to avoid the noted graphical problems of hyperbolic decline curves and to accurately predict the future performance of a producing well, a nonlinear least-squares technique is often considered. This method does not require any straight line extrapolation for future predictions. The mathematical analysis proceeds as follows. The general hyperbolic decline equation for oil production rate (q) as a function of time (t) can be represented as

$$q(t) = q_0 \left(1 + mD_0 t\right)^{-1/m}$$

$$0 < m < 1$$

where

$$q(t) = \text{oil production at time } t$$

$$q_0 = \text{initial oil production}$$

$$D_0 = \text{initial decline}$$

$$m = \text{decline exponent}$$

Also, the cumulative oil production at time t, $Q(t)$, can be written as

$$Q(t) = \frac{q_0}{(m-1)D_0}\left[(1 + mD_0 t)^{\frac{m-1}{m}} - 1\right]$$

where $Q(t)$ = cumulative production as of time t. By combining these equations and performing some algebraic manipulations, it can be shown that

$$q(t)^{1-m} = q_0^{1-m} + (m-1)D_0 q_0^{-m} Q(t)$$

which shows that the production at time t is a nonlinear function of its cumulative production level. By rewriting the equations in terms of cumulative production, we have

$$Q(t) = \frac{q_0}{(1-m)D_0} + q(t)^{1-m}\frac{q_0^m}{(m-1)D_0}$$

It can be seen that the model can be investigated in terms of conventional learning curve techniques, forgetting decline curve, and half-life analysis in a procedure similar to techniques presented earlier in this chapter. The forgetting function has the same basic form as the standard learning curve model, except that the forgetting rate will be negative, indicating a decay process.

The profile of the forgetting curve and its mode of occurrence can influence the half-life measure. This is further evidence that the computation of half-life can help distinguish between learning curves, particularly if a forgetting component is involved. The combination of the learning and forgetting functions presents a more realistic picture of what actually occurs in a learning process. The combination is not necessarily as simple as resolving two curves to obtain a resultant curve. The resolution may particularly be complex in the case of intermittent periods of forgetting.

REFLECTIONS ON LEARNING IN INNOVATION ENVIRONMENT

Degradation of performance occurs naturally, due to either internal processes or externally imposed events, such as extended production breaks. For productivity assessment purposes, it may be of interest to determine the length of time it takes a production metric to decay to half of its original magnitude. For example, for career planning strategy, one may be interested in how long it takes for skills sets to degrade by half in relation to current technological needs of the workplace. The half-life phenomenon may be due to intrinsic factors, such as forgetting, or due to external factors, such as a shift in labor requirements. Half-life analysis can have applications in intervention programs designed to achieve reinforcement of learning. It can also have applications for assessing the sustainability of skills acquired through training programs. Further research on the theory of half-life of learning curves should be directed to topics such as the following:

- Half-life interpretations
- Training and learning reinforcement program
- Forgetting intervention and sustainability programs

In addition to the predictive benefits of half-life expressions, they also reveal the ad-hoc nature of some of the classical learning curve models that have been presented in the literature. We recommend that future efforts to develop learning curve models should also attempt to develop the corresponding half-life expressions to provide full operating characteristics of the models. Readers are encouraged to explore half-life analysis of other learning curve models not covered in this chapter. In some cases, a lower bound is incorporated into the conventional WLC, such that WLC could be represented as:

$$C(x) = \begin{cases} C_1 x^{-b}, & \text{if } x < x_s \\ C_s, & \text{otherwise} \end{cases}$$

where x_s is the number of units required to reach standard cost C_s. Now, if we assume that for some $x_0 < x_s$, the half-life expression becomes:

$$x_{1/2} = \left(\frac{1}{2}\right)^{-\frac{1}{b}} x_0 > x_s$$

What would this mean in an operational context, particularly in dynamic science, technology, and engineering applications? Much research centered on life data needs to be done in this area. The half-life theory approach opens the door to many similar learning curve research inquiries.

MEASURING AGILITY OF INNOVATION

Sieger et al. (2000) present quantitative methodology suitable for adapting to measure the agility of innovation. Achievement of lower product cycle times requires the use of a multidimensional agility measure. In addition to quantifying a manufacturer's ability to respond quickly and cost-effectively to sudden unexpected changes in customer demands, this measure should be able to guide the concurrent engineering effort. Such a measure is described in this chapter. This model utilizes state variable–based performance metrics, which account for the hierarchy among design activities. Also, to address the subjectivity that is inherent in this process and provide confidence interval estimates for the resulting agility measures, simulation modeling is used.

An increasing number of companies are realizing that there is a need to be able to respond quickly, in a cost-effective way, to sudden unexpected changes in customer demands. This more responsive approach results from a competitive-driven trend toward mass customization, as opposed to mass production, and is referred to as agility [5]. One technique that is being used to achieve improved agility levels is to form temporal alliances with outside companies [3]. In this scenario, members of the alliance would emphasize their strengths while collectively minimizing each other's weaknesses. However, as competitors adopt similar temporal alliance strategies, the gains provided by this agile approach would be diminished.

In the manufacturing sector, another method that is also being used to address agility is virtual reality [7]. Although this advanced visualization technique is still in its infancy and currently has substantial capital investment requirements, this will undoubtedly change in the near future. Consequently, as in the previous case, competitive advantages that might be realized by this approach will be reduced as its use becomes widespread.

Ultimately, the responsiveness of companies and/or alliances must be measured in terms of factors that capture the status of the products being developed, relative to their development cycle time. One such approach, which stems from the use of a quantitative performance metric [10], is presented in this chapter. The resulting agility measure extends the existing analysis procedure performed during product development by providing a means for quantifying the actions taken during this process. Using this agility measure as a guidance mechanism, concurrent engineering efforts can be streamlined, minimizing the "concept to cash" time [3].

Development of the required agility measure involves identification of a generic state variable set that captures the essence of the product during its development, association of an implicit hierarchy among these variables, introduction of associated vagueness and loss representations, as well as revisions to the standard design process procedures to accommodate quantification throughout the design portion of this process. Finally, given the process complexity and inherent subjectivity associated with variable representation, simulation is used as a tool to coordinate activity revision policies. Given the difficulty of developing mathematical models that can provide exact analytical solutions for such systems, simulation must be used as a means for numerical analysis [4].

AGILITY MEASURE DEVELOPMENT

The first step in the construction of a framework for agility is the identification of a state variable set. The selection of variables comprising this set is based on the significance of each in relating some aspect of the product's status during its development. The resulting set is denoted by S, and is used to represent the state of the product.

$$S = \left(s_1, s_2, \ldots, s_n\right) \tag{7.1}$$

where the s_i, $i = 1, 2, \ldots, n$ are the key variables that assist in relating the product state. Having identified S, development actions can be applied to transform the current product state, S_C, to a resulting product state, S_R, according to some functional mapping, $f()$.

$$S_R = f\left(x \mid S_C\right) + \varepsilon \tag{7.2}$$

where ε represents the error of the resulting functional approximation.

Although the development actions that are to be implemented are dependent on the product type, knowledge of what actions need to be taken over the course of product development is generally not lacking. In addition, the relative dependence of these actions would typically be evident, to the extent that they could be laid out using a network flow approach; a step that is necessary when employing either the critical path method (CPM) or Program Evaluation Review Technique (PERT) for project analysis. However, while the connectivity among tasks is easily obtained, the design portion of this process, in particular, differs from most projects that commonly use this type of analysis because of its inherently iterative nature. This characteristic of the design process makes such techniques difficult to use directly. As a result, it is much more common to see the design process represented as a systematic procedure using some type of flow chart (e.g., [8, 11]), as opposed to a network flow (e.g., [1]). Despite this difficulty, instead of abandoning the proven benefits provided by network flow analysis in the management of projects, the method presented augments it by considering the two parallel operating planes that are involved. In addition to accounting for the time and resource factors considered by traditional network analysis, and represented in the *product activity plane*, an associated *product state variable plane* is derived. The purpose of this plane is to take into account all performance components that comprise the product state, S. Specifically, there are two basic state variable forms. These two forms are those that result directly from tasks which are performed in the product activity plane, and those that are calculated. The former type are represented by circles and have a one-to-one correspondence with network-level activities. The latter type are indicated by circles having a black interior and are not directly connected to any development activities.

Since agility measures the ability of an organization to respond quickly and cost-effectively to unexpected changes in customer desires, the factors considered in both planes of agility measurement must be aggregated. To do this, a performance metric, $f_P(t)$, is utilized, which provides an overall quantitative assessment of the state variables contained in the product state variable plane.

$$f_P(t) = \frac{L(t) + V(t)}{2} \qquad (7.3)$$

The two terms comprising $f_P(t)$ represent the primary forms of imprecision that are present during product development [12]. In this model, each term has a normalized representation. The first term, $L(t)$, considers the sum of average class losses associated with current development operating points, relative to those that are desired, and has the following form.

$$L(t) = \frac{1}{p} \sum_i avg_j (e^{-a_i L_j(t)}), \quad i = 1, 2, \ldots, p \quad j = 1, 2, \ldots, q$$

where $\qquad (7.4)$

$$a_i > 0 \ \forall i$$

$$q \le n \,(\text{from Eq. 7.1})$$

In this term, the a_is are referred to as product stage multipliers and correspond to specific segments of the development process. The L_is are loss functions, which account for all q state variables having targets. As in the Taguchi representation, three target classes are considered – nominal, large, and small target classes. The second term, $V(t)$, accounts for the average amount of imprecision contained in the r state variables ($r \le n$; also from Eq. 7.1), some of which are fuzzy numbers.

$$V(t) = avg_i \sum_k \sum_x \frac{e^{\mu_k(t,x)} - 1}{rN(e-1)} \qquad (7.5)$$

where i and k account for the p process segments and r state variables being considered, respectively. The membership function, $\mu\,()$, models the state variable values and is numerically evaluated by taking N samples.

Transforming $f_P(t)$ to the lower plane of assessment induces the time structure associated with the task hierarchy. This results in an aggregate performance rating, $f_A(t)$, which is based on the tasks that have been applied. Hence, $f_A(t)$ represents an activity-based performance measure that incorporates all instantaneous performance ratings resulting from the activities that were performed up to the current operating point, t.

$$f_A(t) = \frac{1}{t} \int_{\tau=0}^{t} f_P(\tau) d\tau \qquad (7.6)$$

The lower limit on the integral for $f_A(t)$ represents conceptualization of the product, and the upper limit goes to delivery of the product to market. Since this range addresses "concept to cash," this function can be used to provide the agility metric, A, given here:

$$A = \max\{e^{-0.001c_T t} f_A(t)\} \tag{7.7}$$

where c_T represents an aggression factor, which is used to adjust the expected rate of convergence to a specific point in the process. As an example, this factor could be set to converge at the end of the conceptual design stage; the rationale being that this portion of the development process should be emphasized, because approximately 80% of the development cost is typically committed by its conclusion [11]. The respective aggressive action factors corresponding to A_1, A_2, A_3, A_4, and A_5 are 0.1, 1, 10, 100, and 1,000. Therefore, by increasing the value of c_T (i.e., becoming more aggressive), the expectation for product clarity becomes more heavily weighted toward the early stages of product development. For example, in routine design, where the domain knowledge and its application are well known, there would likely be a greater emphasis placed on early solution convergence. In contrast, it is expected that for creative design problems, the c_T value would be relaxed to allow for improved performance levels later in the development process. The relationship between $f_P(t)$, $f_A(t)$, and A can be viewed graphically (Sieger et al., 2000). The agility metric tends toward a rating of 1.0, as the rise time of the activity-based performance measure is reduced. Since $f_A(t)$ integrates the process performance values provided by $f_P(t)$, A provides the desired time-based performance rating, which reflects the agility of a manufacturer during product development.

MODIFICATIONS TO DESIGN PROCESS PROCEDURES

To develop a quantitative performance metric that can be used throughout the product development process, and subsequently provide a measure for agility, it is necessary to modify the current design practices performed during the first two stages of the design process (i.e., during the specification and planning and the conceptual design stages). In this section, the modifications that were made to these standard procedures are summarized.

Specification and planning stage modifications transform the standard six-step Quality Function Deployment (QFD) model into a seven-step procedure referred to as the Enhanced QFD model (EQFD). This EQFD model differs from Clausing's version in that the emphasis here is on quantification, while Clausing stresses piece-part decomposition [2]. In applying the standard QFD approach, the six steps that can be implemented are solicitation of customer requirements (step 1), assignment of relative weights to customer requirements (step 2), competition benchmarking (step 3), selection of a list of engineering requirements that address the known customer requirements (step 4), assessment of engineering requirement correlations (step 5, not always performed), and determination of engineering targets (step 6).

As in the QFD model, the EQFD procedure obtains a list of customer requirements as its first step. However, all the remaining steps have been modified. In these steps, EQFD evaluation involves using a fuzzy rank-order approach to assign relative weights to customer requirements (step 2), establishing a customer requirement fulfillment rating based on expectation (step 2′), implementing a multivalent approach for competition benchmarking (step 3), utilizing a fuzzy "better/worse" technique in determining engineering/customer requirement relation weights (step 4, adapted from [6]), calculating an engineering/customer requirement expectation rating (step 4′), and assigning fuzzy measures for the resulting engineering targets (step 6). In this approach, an optional step that can weight customers on the basis of marketing information and objectives of the company can also be included (step 0). In addition, step 5 of the QFD procedure need not be performed in the EQFD model, as it is accounted for in the fourth step of the EQFD.

In a similar fashion, the conceptual design stage has also been adapted. In this stage, both the existing and revised procedures start by determining an overall functional objective. This objective is then decomposed into a set of known primitives. In performing this decomposition, the revised approach elicits ratings for customer satisfaction and functional independence; components that are only implicitly known in the standard approach. Next, both approaches generate a list of design concepts. In the existing procedure, bivalent evaluation of feasibility, readiness, and requirement fulfillment is performed. The revised approach implements a fuzzy variant of this procedure. Having undergone preliminary concept evaluation, final concept evaluation is carried out using Pugh's Method [9]. In the revised model, a fuzzy datum technique is utilized to increase the amount of information that is obtained in this step. Finally, the revised model adds two additional steps, which use the fuzzy ratings obtained during preliminary and final concept evaluations. In these steps, this information is used to assess set and supra-set desirability. In this case, the objective is to relate the quality of the concepts from which the designers are selecting.

Given the unique symbolic state representations, causal relationships between development activities and their counterparts in the product state variable plane can be expressed functionally and are given in the following. Although this causality is due to a dependence on other state variables and activities occurring in the product activity plane, functional dependence only explicitly accounts for the state variable dependency. This approach reflects the desire to enforce performance thresholds on product development activities. Provided that acceptable levels of performance have been attained, subsequent tasks can be undertaken. For example, the function corresponding to the assignment of customer weights, f_9, is triggered by its associated activity in the product activity plane but is subject to the achievement of acceptable performance levels on two predecessor activities. These two activities contribute to the generation of state variable values resulting from the formation of the design team, f_4, and identification of the customer, f_5. State variables that are calculated (i.e., not triggered by actions in the product activity plane) are activated as soon as their predecessors become active.

Connectivity among all state variables comprising the product state variable plane is illustrated graphically in Sieger et al. (2000). Of the 38 state variables being considered in that example, 6 are calculated, and 32 result directly from design actions.

Using the modifications described in the previous section facilitates quantification early in the design process but does not eliminate the uncertainty associated with the design components. Consequently, the variability, which results from the inherent subjectivity in specifying exact values for the fuzzy number supports, is modeled by centering uniform probability distributions around each of the designer's point estimates (i.e., the values that have been specified as the supports for the initial trapezoidal fuzzy number representations). Since the designer's flexibility to change parameter values is reduced as the time that has been allocated for product development is spent, the span of these uniform distributions is inversely related to design time.

In simulating the design portion of the product development process, fuzzy numbers are adapted to reflect the improvements in clarity with regard to their referent design parameters. In this approach, the initial triangular fuzzy number form is shown along the vertical axis. As design tasks are performed, the uncertainty associated with this design parameter converges to a specific value. This convergence process is depicted using four lines to track the adaptation of each of the fuzzy trapezoidal supports (i.e., the center support of the initial representation is split to form two supports during the first part of this graph).

To guide selection of design actions, a reference performance rating must be provided. This reference equates to the design team's expectation for performance during product development. Depending on the stage of this process, and whether or not the reference exceeds the actual performance, f_P, design actions may be repeated. Selection of which action to repeat is dependent on the values of specific state variables and their relation to the task hierarchy. Readers interested in this quantitative approach should refer to Sieger et al. (2000) for further details.

Given product state variables, parametric representations, design tasks, task dependence, and a reference performance, product development participants can use simulation as a means for establishing point and interval estimates for agility. Included in the computational results are the f_A, c_T, and A values for each of the ten runs that were performed. The f_A values correspond to the activity-based performance obtained at the end of the design process, and the c_T values represent different convergence policies for agility. From these results, the average and half-width, for the agility metric were determined. If $c_T = 1$ was the selected policy for agility measurement, the results indicate that the agility measure of 0.456 is within 5.5% of the true, but unknown, agility. To obtain tighter bounds on this measure, the number of simulation runs must be increased and/or the variance of state variables must be reduced. However, since the variability in the state variables reflects the capability of the designers to converge to a product that meets customer and engineering requirements, it can only be reduced indirectly. Examples of indirect methods for reducing variability include training, improved communication between designers, inclusion of more experienced designers in the design team, formation of temporal alliances, use of virtual reality, etc. The agility metric that has been described provides a useful means for determining the responsiveness of an organization to sudden unexpected changes in customer demands. This capability of the metric results from a lower-level performance modeling approach, which accounts for state variables associated with product development. By dynamically capturing this information, it is anticipated that concurrent engineering efforts could be streamlined. This improvement over current practices is due to the guidance provided to the design team by the agility and performance measures. Moreover, using simulation, the accuracy of the results provided by this procedure can also be determined and used as a means for establishing certainty. Thus, it is expected that this approach would improve the outputs that would be obtained from the product development process.

CONCLUSIONS

The premise of this chapter is that whatever cannot be quantitatively assessed is difficult to measure. How do we know when innovation has occurred? Is innovation repeatable? Can innovation be replicated elsewhere? These are important questions that utility modeling can help to address, even if they are not conclusively answered. The contents of this chapter provide the quantitative framework for assessing investments in innovation and evaluating the outputs of innovation. In addition to utility modeling, the chapter also presents the technique of polar plots for comparing performance attributes of alternate innovation projects.

REFERENCES

Badiru, Adedeji B. (1991). *Project Management Tools for Engineering and Management Professionals*, Industrial Engineering & Management Press, Norcross, GA.

Badiru, Adedeji B. (1995a). Multivariate Analysis of the Effect of Learning and Forgetting on Product Quality. *International Journal of Production Research*, Vol. 33, No. 3, 1995, pp. 777–794.

Badiru, Adedeji B. (1995b, June). Incorporating Learning Curve Effects into Critical Resource Diagramming. *Project Management Journal*, Vol. 26, No. 2, pp. 38–45.

Badiru, Adedeji B. (1994). Multifactor Learning and Forgetting Models for Productivity and Performance Analysis. *International Journal of Human Factors in Manufacturing*, Vol. 4, No. 1, pp. 37–54.

Badiru, Adedeji B. and Anota Ijaduola (2009, June). Half-Life Theory of Learning Curves for System Performance Analysis. *IEEE Systems Journal*, Vol 3, No 2, pp. 154–165.

Badiru, Adedeji B. (2008) . Triple C *Model of Project Management: Communication, Cooperation, and Coordination*, Taylor & Francis CRC Press, Boca Raton, FL.

Belkaoui, A. (1976). Costing Through Learning. *Cost and Management*, Vol. 50, No. 3, pp. 36–40.

Camm, Jeffrey D., James R. Evans, and Norman K. Womer (1987). The Unit Learning Curve Approximation of Total Cost. *Computers and Industrial Engineering*, Vol. 12, No. 3, pp. 205–213.

Globerson, S., A. Nahumi, and S. Ellis (1998). Rate of Forgetting for Motor and Cognitive Tasks. *International Journal of Cognitive Ergonomics*, Vol. 2, pp. 181–191.

Glover, J. H. (1966). Manufacturing Progress Functions: An Alternative Model and Its Comparison with Existing Functions. *International Journal of Production Research*, Vol. 4, No. 4, pp. 279–300.

Jaber, M. Y. and M. Bonney (1996). Production Breaks and the Learning Curve: The Forgetting Phenomena. *Applied Mathematical Modelling*, Vol. 20, pp. 162–169.

Jaber, M. Y. and M. Bonney (2003). Lot Sizing with Learning and Forgetting in Setups and in Product Quality. *International Journal of Production Economics*, Vol. 83, No. 1, pp. 95–111.

Jaber, M. Y. and M. Bonney (2007). Economic Manufacture Quantity (EMQ) Model With Lot Size Dependent Learning and Forgetting Rates. *International Journal of Production Economics*, Vol. 108, Nos. 1–2, pp. 359–367.

Jaber, M.Y. and A.L. Guiffrida (2004). Learning Curves for Processes Generating Defects Requiring Reworks. *European Journal of Operational Research*, Vol.159, No.3, pp. 663–672.

Jaber, M.Y. and A. L. Guiffrida (2008). Learning Curves for Imperfect Production PROCESSEs with Reworks and Process Restoration Interruptions. *European Journal of Operational Research*, Vol. 189, No. 1, pp. 93–104.

Jaber, M. Y. and S. Sikstrom (2004). A Numerical Comparison of Three Potential Learning and Forgetting Models. *International Journal of Production Economics*, Vol. 92, No. 3, pp. 281–294.

Jaber, M. Y., H. V. Kher, and D. Davis (2003). Countering Forgetting Through Training and Deployment. *International Journal of Production Economics*, Vol. 85, No. 1, pp. 33–46.

Knecht, G. R. (1974, September). Costing, Technological Growth, and Generalized Learning Curves. *Operations Research Quarterly*, Vol. 25, No. 3, pp. 487–491.

Levy, F. K. (1965). Adaptation in the Production Process. *Management Science*, Vol. 11, No. 6, pp. B136–B154.

Liao, W. M. (1979). Effects of Learning on Resource Allocation Decisions. *Decision Sciences*, Vol. 10, pp. 116–125.

Mazur, J. E. and R. Hastie (1978). Learning as Accumulation: A Reexamination of the Learning Curve. *Psychological Bulletin*, Vol. 85, pp. 1256–1274.

McIntyre, E. V. (1977). Cost-Volume-Profit Analysis Adjusted for Learning. *Management Science*, Vol. 24, No. 2, pp. 149–160.

Nanda, Ravinder (1979). Using Learning Curves in Integration of Production Resources. *Proceedings of 1979 IIE Fall Conference*, pp. 376–380.

Nembhard, D. A. and N. Osothsilp (2001). An Empirical Comparison of Forgetting Models. *IEEE Transactions on Engineering Management*, Vol. 48, pp. 283–291.

Nembhard, D. A. and M. V. Uzumeri (2000). Experiential Learning and Forgetting for Manual and Cognitive Tasks. *International Journal of Industrial Ergonomics*, Vol. 25, pp. 315–326.

Park, Chan S. and Gunter P. Sharp-Bette (1990). *Advanced Engineering Economics*, Wiley & Sons, New York.

Pegels, Carl C. (1976, October). Start Up or Learning Curves - Some New Approaches. *Decision Sciences*, Vol. 7, No. 4, pp. 705–713.

Richardson, Wallace J. (1978). Use of Learning Curves to Set Goals and Monitor Progress in Cost Reduction Programs. *Proceedings of 1978 IIE Spring Conference*, pp. 235–239.

Sieger, David, A. B. Badiru, and M. Milatovic (2000). A Metric for Agility Measurement in Product Development. *IIE Transactions*, Vol. 32, pp. 637–645.

Smith, Jason (1989). *Learning Curve for Cost Control*, Industrial Engineering and Management Press, Norcross, GA.

Smunt, Timothy L. (1986). A Comparison of Learning Curve Analysis and Moving Average Ratio Analysis for Detailed Operational Planning. *Decision Sciences*, Vol. 17, pp. 12–19.

Sule, D. R. (1978). The Effect of Alternate Periods of Learning and Forgetting on Economic Manufacturing Quantity. *AIIE Transactions*, Vol. 10, No. 3, pp. 338–343.

Towill, D. R. and U. Kaloo (1978). Productivity Drift in Extended Learning Curves. *Omega*, Vol. 6, No. 4, pp. 295–304.

Womer, N. K. (1979). Learning Curves, Production Rate, and Program Costs. *Management Science*, Vol. 25, No. 4, pp. 312–219.

Womer, N. K. (1981). Some Propositions on Cost Functions. *Southern Economic Journal*, Vol. 47, pp. 1111–1119.

Womer, N. K. (1984). Estimating Learning Curves from Aggregate Monthly Data. *Management Science*, Vol. 30, No. 8, pp. 982–992.

Womer, N. K. and T. R. Gulledge Jr (1983). A Dynamic Cost Function for an Airframe Production Program. *Engineering Costs and Production Economics*, Vol. 7, pp. 213–227.

Wright, T. P. (1936, February). Factors Affecting the Cost of Airplanes. *Journal of Aeronautical Science*, Vol. 3, No. 2, pp. 122–128.

Yelle, Louis E. (1976, June). Estimating Learning Curves for Potential Products. *Industrial Marketing Management*, Vol. 5, No. 2/3, pp. 147–154.

Yelle, Louis E. (1979, April). The Learning Curve: Historical Review and Comprehensive Survey. *Decision Sciences*, Vol. 10, No. 2, pp. 302–328.

Yelle, Louis E. (1983, December). Adding Life Cycles to Learning Curves. *Long Range Planning*, Vol. 16, No. 6, pp. 82–87.

8 Innovation on the Edge of Dollars
A Case Study

INTRODUCTION

As a testament to the theme of this book, this chapter presents a research case study published by a team at the US Air Force Institute of Technology in 2022 (Ryan et al, 2022). The cautionary note conveyed by the title of the published research is that "practice doesn't make perfect." Just because we have innovation doesn't mean we get profitable dollars from the venture. Badiru and Barlow (2019) emphasize the role of innovation in national defense strategies. Badiru (2020) addresses the same topic from a systems perspective. Badiru and Lamont (2022) present innovation implementations from an integrated quantitative and qualitative framework. Using a diverse literature background, Ryan et al. (2022) conducted a case study that ties innovation to business dollars.

RESEARCH BACKGROUND

The authors of this research (Ryan et al., 2022) examine and evaluate organizational factors associated with commercialization under the Air Force Small Business Innovation Research (SBIR) program. Their objective is to improve return on investment. The data set used was the SBIR Phase II program data set, which contains information on 433 SBIR topics with closed contracts reported during Department of Defense (DoD) fiscal years (FYs) 2015 to 2018. Each data point contained characteristics of the topic, including commercialization. Military capability or topic areas were hypothesized to have varying commercialization rates. Incumbency was theorized to be a characteristic of successful programs, while increased company size was theorized as a characteristic of unsuccessful programs. Variables were analyzed through graphs and logistic regression.

Small businesses (1 to 31 employees) have a 2.6% increased commercialization rate compared with large businesses (32 to 499 employees); this increase is significant when compared with the 8.8% global success rate of SBIR projects. No learning effect or improved performance was observed between companies new to the SBIR program (fewer than 14 contracts) and incumbents (15,419 contracts). The opposite learning was observed with new entrants outperforming incumbents. A bump in the data appears for newer entrants with some experience.

In FY 2019, DoD obligated $1.8 billion in SBIR funding, and previous research indicated that the commercialization rate of SBIR Phase II contracts is approximately

           DOI: 10.1201/9781003403548-8

8.8%. This exploratory research looks at factors and trends seen in successful programs. Findings indicate factors that may guide investment choices to improve commercialization rates.

This research focuses on the performance of SBIR investments in defense-related technologies. Understanding the performance of SBIR investments can provide insight into better investment strategies and thus, more effective interaction with the commercial sector. The National Defense Strategy recognizes that many technological developments will come from the commercial sector (Mattis, 2018). Innovation has the potential to drive economic growth and international competitiveness (Balzat, 2006). While innovation involves the generation, adoption, implementation, and incorporation of new ideas, practices, and artifacts (Van de Ven and Poole, 1989), our measure of performance considers the actual commercialization of innovation beyond early investment.

Our ability to innovate effectively has strategic importance. The national security of the USA depends on the ability to gain access to and make the best use of innovations. The 2018 National Defense Strategy (Mattis, 2018) highlights this role of innovation with the following quote:

> Success no longer goes to the country that develops a new technology first, but rather to the one that better integrates it and adapts its way of fighting.
>
> James Mattis, USA Secretary of Defense

Regardless of strategic focus, whether international terrorism or the rival powers of Russia and China, our ability to develop and infuse innovation is crucial to our nation's defense. While internal investments (e.g., Air Force Research Laboratory) are important to developing defense-focused technologies, our ability to foster and leverage innovation in our industrial base is vital. The Department of Defense (DoD) faces the challenges of attracting these external innovators and bringing their ideas to fruition in a way that enhances the capability of the armed forces. One of the many ways the DoD attempts to accomplish this external investment is through the SBIR program, which is a federal government program that deliberately invests research money in small businesses. The Small Business Administration (SBA) started the SBIR program in 1977 to support innovation through the investment of federal research funds in critical American priorities to build a strong national economy. The SBA (n.d.) explains how the program was established under the Small Business Innovation Development Act of 1982 (Small Business Innovation Development Act, 1982), with the purpose of strengthening the role of innovative small business concerns in federally funded research and development (R&D). Through a competitive awards-based program, SBIR allows "small businesses to explore their technological potential and provides the incentive to profit from its commercialization" (SBIR | STTR, n.d., para. 1). Beyond the critical technologies and access to external innovators, SBIR investments serve as an economic stimulus to strengthen the industrial base. Known as "America's Seed Fund," SBIR works to stimulate high-tech innovation in the United States while targeting specific research and development needs of the government (SBIR | STTR, n.d.). SBIR is one of the largest DoD-backed innovation programs in operation. In FY 2019, the DoD obligated $1.8 billion in SBIR funding. SBIR investments target a specific segment of innovators within the domestic economy – small businesses.

Traditionally, the Air Force has followed a "pull" model of innovation with SBIR investments by broadcasting its needs to participating small businesses. These needs are based on topics generated throughout the Air Force. Capability needs (i.e., SBIR topics) are published, and small businesses reply with proposals. Accepted proposals, regardless of sponsor, follow a three-phase program. Table 8.1 provides descriptions of the phases, along with their funding and timing. To participate in the SBIR program, firms must be eligible, have an adequate plan to accomplish the required research, and conduct the research within the USA. Eligibility is restricted to businesses with 500 or fewer employees and is established on initial application as well as through certifications at other times during participation. Participating firms must also provide plans to meet research requirements for Phases I and II. The research must be done in the USA unless the funding agreement officer recognizes a unique circumstance that demands otherwise. If the small business qualifies, then the business will be eligible to participate.

This research considered programs that met the basic eligibility and planning for Phase I; additionally, researchers met more rigorous requirements established for Phase II. To secure Phase II award, all programs developed commercialization plans. Elements of SBIR commercialization plans can include company information, customer data, data on competition, market assessments, data regarding intellectual property, and financing. Further, award of Phase II requires the submittal of a business plan, executive summary, cost proposal, and technical proposal. This documentation undergoes a rigorous review process to ensure that only the most meritorious scientific proposals are funded (Kelly and Sensenig, 2019).

An SBIR project is considered successful when the product is commercialized. Commercialization occurs when a project progresses beyond seed funding through

TABLE 8.1

Phases of SBIR Programs

Phase	Objective	Funding	Period of Performance
Phase I	Establish technical merit, feasibility, and commercial potential; complete at least one-third of required research	<$150,000 (SBIR)	6 months
Phase II	Assess scientific and technical merit and commercial potential; complete an additional half of the required research for the program	<$1,000,000 (SBIR)	24 months
Phase III	Commercialization	Other sources	N/A

SBIR to longer-term governmental or commercial funding (SBIR | STTR, n.d.). Transition into Phase III represents this commercialization; programs in Phase III transition into the broader Service branches or agencies that need them (Bresler, 2018). Air Force SBIR programs from 2015 to 2018, which represent our data set, had a Phase II to Phase III transition rate of 8.8% (Blake, 2020; Rask, 2019).

Considering the degree of need for DoD investment and innovation throughout the nation, understanding the factors that can influence the success or failure of these programs is valuable. Success in SBIR programs occurs when the programs transition from government seed funds to external funds or non-SBIR funds, whether governmental or commercial. While ideation and prototyping are outputs of this process, innovation is considered successful when the invention is implemented and adopted (Fagerberg and Mowery, 2006). This transition, from seed funds to external funds, is defined as commercialization, and it is the accepted measure of success for SBIR programs (SBIR | STTR, n.d.).

This research analyzes 433 Air Force SBIR projects from 2015 to 2018 to discern factors related to their transition success. This set includes only programs that have successfully demonstrated technical feasibility and have completed a contracted R&D phase (e.g., Phase I and II completed). This three-year baseline represents a time of relative stability, before the more recent phase of experimentation witnessed with AFWERX and other organizations. The stability of this baseline allows a factor analysis across this broad set of projects; it also enables a stable point of comparison for recent efforts. Air Force SBIR projects were selected due to availability of data and sponsorship by the Air Force SBIR office.

We consider two levels of analysis: the entire portfolio and capability-based segments. Air Force investments are diverse, ranging from landing gear corrosion prevention to artificial intelligence algorithms to bolster battlespace awareness. Segmentation permits a more nuanced comparison of investments and transition both within the portfolio and across military capability. Capability-based portfolio segmentation was accomplished in previous research (Rask, 2019) and leveraged the well-established Joint Capability Area (JCA) taxonomy, which provides common language for DoD capabilities (Joint Chiefs of Staff, 2018). Our focus is on factors that are known pre-award: What factors can we know in advance of award that may influence the decision of the Air Force SBIR office? Based on what we can know, can we make choices that improve our success? Previous analysis found a commercialization rate of 8.8% (Blake, 2020; Rask, 2019). Considering the number of projects and investments, small improvements matter in this space. As an example, achieving a transition rate of 10% represents five additional capabilities transitioning to use. If factors that correlate with success can be determined, policy can be shaped to target improvements and increase the commercialization rate for our SBIR investments.

Two independent factors are considered: the size (number of employees) of the small business and experience (history of working with the government). The primary funding of this article relates to small business size; smaller businesses have a statistically significant transition advantage over their larger counterparts. Firms with 31 or fewer employees ($n = 217$) had a transition rate 2.6% higher than firms with 32 to 499 employees ($n = 215$). Our second finding relates to experience: no evidence supports a hypothesis that experience working with the government improves a firm's transition performance. Firms with an average of five contracts with the government ($n = 217$ contracts) had a significant improvement in performance (commercialization) when compared with those with an average of 73 or more contracts with the government ($n = 215$ contracts).

Data Set

This research analyzes 433 Air Force SBIR projects from 2015 to 2018. This set includes only programs that both successfully demonstrated technical feasibility and completed a contracted research and development phase (e.g., Phases I and II completed). Further, the set considers only programs that reached the point of transition to Phase III and included programs that were either commercialized or not.

The three-year baseline from 2015 to 2018 represents a time of relative stability. More recent innovation efforts have witnessed experimentation in investment strategies (e.g., AFWERX). The stability of this baseline allows a factor analysis across this broad set of projects; it also provides a stable point of comparison for recent experimentation efforts. Additionally, since these data are less than ten years old, relevant follow-on research, as needed, is facilitated. Ten years is considered recent enough to preserve accurate memories of key informants in the event that follow-on interviews or interaction are required.

This population provided for a consideration of the performance of external investments across a broad range of military capabilities and technologies. Consistent trends across the set and within capabilities permit generalization of the results beyond idiosyncrasies that may be present in certain technologies. The distribution of these investments across areas of military capability and their relative success are shown in Table 8.2.

Factors Considered

In addition to a project's commercialization (our dependent variable), we sought factors that are known in advance of investment. Analysis of ex ante factors may reveal trends that can enable prediction and inform investment strategies. Three of these factors were considered: (a) military capability area pursued (control variable), (b) historical firm engagement with the government (independent variable), and (c) firm size (independent variable). These areas were chosen due to data availability, qualitative observations of the data set, and theories from innovation research. Our unit of analysis is individual SBIR topics. An SBIR topic is a description of need that is released to prospective innovators for their subsequent bids. The topics spanned technologies from novel anticorrosion coatings to global satellite command and control systems. Due to this diversity, a means to segment

TABLE 8.2

SBIR Investments across Military Capabilities

Joint Capability Area	Number of Investments	Percentage Successful
Force Support. The ability to establish, develop, and maintain a mission-ready Joint Force and build relationships with foreign and domestic partners.	9	22%
Battlespace Awareness. The ability to understand dispositions and intentions as well as the characteristics and conditions of the operational environment that bear on national and military decision-making by leveraging all sources of information, including intelligence, surveillance, reconnaissance, meteorological, and oceanographic.	74	12%
Force Application. The ability to integrate maneuver and kinetic, electromagnetic, and informational freedom to gain a position of advantage and/or create lethal or nonlethal effects on designated targets.	82	7%
Logistics. The ability to project and sustain the Joint Force.	78	4%
Command and Control. The ability to exercise authority and direction by a properly designated commander or decision-maker over assigned and attached forces and resources in the accomplishment of the mission.	7	0%
Communication and Computers. The ability to exercise authority and direction by a properly designated commander or decision-maker over assigned and attached forces and resources in the accomplishment of the mission.	73	8%
Protection. The ability to preserve the effectiveness and survivability of military and nonmilitary personnel, equipment, facilities, and infrastructure by preventing, mitigating, and ensuring recovery from attacks, chemical, biological, radiological, and nuclear (CBRN) incidents, and other hazards.	19	10%
Corporate Management and Support. The ability to provide strategic senior-level, enterprise-wide leadership, direction, coordination, and oversight through a chief management officer function.	90	11%

the portfolio for analysis was sought. Segmentation allows cross-portfolio and within-segment analysis.

Previous research of this data set categorized each SBIR project based on the military capability area it satisfied (Rask, 2019). The Joint Staff's JCA listing was used for this purpose (Joint Chiefs of Staff, 2018). This choice of an existing, defense-related taxonomy facilitates analysis focused on specific areas of military need. The choice of capability-based segmentation blends two factors – technology and market segment for application. Certain capabilities rely on a limited set of technologies.

Further, patterns of success and failure could be due to the maturity or market associated with a capability area. Where the force application capability area is uniquely
military, communications and computers has a wide range of applications and could
potentially represent a thriving commercial innovation base.

Our next two factors, incumbency and size, shift our attention from the technology being sought to characteristics of the firms completing the work. Incumbency
is a measure of historic interaction with the government. We operationalize incumbency as the number of government contracts held by a firm. Contracting with the
government introduces complexities for small firms (Schilling et al., 2017). We
hypothesize that increased experience working with the government reduces these
challenges; through iteration, a firm learns government processes and needs. As an
extension, we assume that experience with the government should improve the probability of commercialization.

The size of a firm can have multiple effects on performance. Literature on innovation with the government points to administrative burdens that do not favor smaller
firms (Schilling et al., 2017). However, innovation literature has observed higher performance in smaller and fatter organizations (Quinn, 1985). The larger an organization becomes, the more likely it is to develop a hierarchical structure that may reduce
innovation performance (Kirsner, 2018). Further, with increased organizational size,
"effectiveness of internal knowledge flow dramatically diminishes and degree of
intra-organizational knowledge sharing decreases" (Serenko et al., 2007, p. 614). We
hypothesize that smaller companies will perform better than larger companies, yet
what small and large represent is not certain.

METHODOLOGY

The objective of this research is to understand factors that are correlated to SBIR
project success with the aim of improved investments. The data set we used was the
SBIR Phase II program data set, which contains information on 433 SBIR topics
with closed contracts reported during DoD FYs 2015 to 2018.

ANALYSIS

Two methods were used to analyze this data set: logistic regression and hypothesis
testing associated with population comparisons. The first method, logistic regression, was selected due to the binary characteristic of the dependent variable (e.g.,
whether or not a project transition occurred). This analysis technique can provide
a probability of success as a function of independent variables (company size and
recidivism). Military capability areas were included as control variables. These military capabilities were assigned as part of previous research in which a panel of raters
categorized each project into one of eight joint capability areas (Rask, 2019). We
did not find a statistically significant relationship between transition success and the
independent variables. Using an open-source development environment for statistical analysis, R Studio, the probability of commercialization was estimated by fitting

a logistic regression model. A summary of the results from this model is reported in Table 8.3.

This lack of correlation may be due to the lack of an effect. However, it may also be due to the variation within the data set even following segmentation. As mentioned earlier, the capability-based segmentation has at least two factors within it – technology and market. The set may still be too noisy with too many effects to discern a relationship. Our second analysis method, hypothesis testing using population comparison, is a coarser analysis, allowing a binary result. Are commercialization rates of populations the same or different, and if different, to what extent? For example, in Table 8.4, commercialization rates (success rates) of quartiles are not the same. This technique is more resilient to noise in the data; however, it does not provide a relationship between the variables.

We have made comparisons of subpopulations within the set, determining whether commercialization in those populations is significantly different. Two separate analyses were completed with the data based on the independent variables of recidivism and company size. In both analyses, the performance of the upper and lower quartiles as well as the upper and lower halves of the sets were compared to determine whether a difference existed. The data ranged from companies with no previous government interaction to companies with over 400 SBIR contracts awarded. The set was broken into nearly even quartiles, and hypothesis testing was performed to compare the upper and lower quartiles (new entrants and experienced firms). This hypothesis testing was repeated with the set broken into two nearly even halves. The average number of contracts awarded was 39.

Table 8.4 provides the quartiles and halves and success rates for the variable. The lower quartile ranged from 1 to 4 awards (111 firms), while the upper quartile ranged from 36 to 419 awards (106 firms). The average success rates were 7.2% for the lower quartile and 7.5% for the upper quartile. No significant difference exists between new entrants and incumbent firms ($P = 0.10$). The lower half ranged from 1 to 14

TABLE 8.3
Logistic Regression Model Results

Variable	Coefficient	P Value	Average Marginal Effect
Number of Employees	−0.000714	0.760	−0.0001
Total_Awards	−0.004566	0.326	−0.0004
JCA_1	1.055896	0.247	0.0823
JCA_2	0.101916	0.836	0.0079
JCA_3	−0.383365	0.483	−0.0299
JCA_4	−0.871345	0.206	−0.0679
JCA_5	−14.250346	0.987	−1.1101
JCA_6	−0.299243	0.584	−0.0233
JCA_7	−0.063236	0.939	−0.0049

P-values of 0.05 or less indicate significant results, which were not found in the set. JCA = Joint Capability Area.

TABLE 8.4
Repetition

Quartile	# Awards	# Commercialized	Success Rate	Quartile Size
1	≤4	8	7.2%	111
2	5–14	14	13.2%	106
3	15–35	8	7.3%	109
4	36–419	8	7.5%	106
Half	**# Awards**	**# Commercialized**	**Success Rate**	**Half Size**
1	≤14	22	10.1%	217
2	15–419	16	7.4%	215

awards (217 firms), while the upper half ranged from 15 to 419 awards (215 firms). The average success rates were 10.1% for the lower half and 7.4% for the upper half. A statistically significant difference does exist between new entrants and incumbent firms ($P = 0.10$). We expected that experienced companies would outperform new entrants. However, it appears that no clear learning or improved performance occurs as companies repeatedly interact with the SBIR program. Surprisingly, new entrants appear to have improved performance. Of interest, a bump in the data appears when comparing the success rate of the second quartile with other quartiles. The second quartile ranged from 5 to 14 awards (106 firms) with a success rate of 13.2%. This is a statistically significant result when compared with all other quartiles ($P = 0.09$). It appears that newer entrants with some experience perform higher than either new entrants or incumbents with more extensive experience. The increase from the first to second quartile may be due to learning; however, what dynamics explain the drop in performance?

Next, the population of projects was segmented based on size, where the lower quartile (companies with 1 to 14 employees) was compared with the upper quartile (from 96 to 500). The small companies did not have a statistically significant difference in performance from the larger companies. The lower half (companies with 1 to 31 employees) was compared with the upper half (32 to 499 employees). The success rates for both quartiles and halves are found in Table 8.5. The small companies

TABLE 8.5
Company Size

Quartile	# Employees	# Commercialized	Success Rate	Quartile Size
1	≤14	11	9.4%	117
2	15–31	11	11.0%	100
3	32–95	7	6.5%	108
4	96–499	9	8.4%	107
Half	**# Employees**	**# Commercialized**	**Success Rate**	**Half Size**
1	≤31	22	10.1%	217
2	32–499	16	7.4%	215

TABLE 8.6

Battlespace Awareness

Quartile	# Awards	# Commercialized	Success Rate	Quartile Size
1	≤5	3	15.8%	19
2	6–16	3	15.8%	19
3	17–31	2	11.1%	18
4	32–419	1	5.6%	18
Quartile	# Employees	# Commercialized	Success Rate	Quartile Size
1	≤15	3	15.8%	19
2	16–33	3	16.7%	18
3	34–78	2	10.5%	19
4	79–334	1	5.6%	18

had a commercialization rate of 10.1%, whereas the larger companies had a rate of 7.4%. This difference was statistically significant ($P = 0.10$). Consistently with the literature, we find that smaller companies performed better than larger companies.

For both independent variables, we then considered performance within large portfolio categories. Comparisons between quartiles were performed (Tables 8.6 to 8.9); however, there were not enough data points to yield a statistically significant result. We are able to draw conclusions only based on the entire population and not the segments.

Finally, we examined the intersection of the two independent variables – recidivism and company size. The average company size of each recidivism quartile was determined (Table 8.10). Additionally, we determined the average number of awards for each company size quartile. It appears that newer entrants are also, on average, smaller companies. Alternatively, it appears that larger companies are, on average, the incumbent.

TABLE 8.7

Force Application

Quartile	# Awards	# Commercialized	Success Rate	Quartile Size
1	≤5	3	13.0%	23
2	6–13	2	11.1%	18
3	14–28	1	4.5%	22
4	29–419	0	0.0%	19
Quartile	# Employees	# Commercialized	Success Rate	Quartile Size
1	≤14	3	15.0%	20
2	15–25	2	9.5%	21
3	26–69	1	4.8%	21
4	70–482	0	0.0%	20

TABLE 8.8
Communication and Computers

Quartile	# Awards	# Commercialized	Success Rate	Quartile Size
1	≤5	2	10.0%	20
2	6–14	1	5.6%	18
3	15–47	1	5.9%	17
4	48–419	2	11.1%	18
Quartile	# Employees	# Commercialized	Success Rate	Quartile Size
1	≤15	3	14.3%	21
2	16–30	1	6.3%	16
3	31–110	1	5.3%	19
4	111–334	1	5.9%	17

TABLE 8.9
Corporate Management and Support

Quartile	# Awards	# Commercialized	Success Rate	Quartile Size
1	≤3	2	10.0%	20
2	4–6	2	13.0%	23
3	7–31	4	12.0%	25
4	32–151	2	9.1%	22
Quartile	# Employees	# Commercialized	Success Rate	Quartile Size
1	≤11	0	0.0%	23
2	12–34	6	27.3%	22
3	35–85	1	4.3%	23
4	86–494	3	13.6%	22

TABLE 8.10
Company Recidivism and Size

Quartile	# Awards	Average Company Size	Success Rate	Quartile Size
1	≤4	32	7.2%	111
2	5–14	43	13.2%	106
3	15–35	73	7.3%	109
4	36–419	125	7.5%	106
Quartile	# Employees	Average # Awards	Success Rate	Quartile Size
1	≤14	6	9.4%	117
2	15–31	17	11.0%	100
3	32–95	41	6.5%	108
4	96–499	100	8.4%	107

DISCUSSION OF RESULTS

Our results focus on patterns with the two independent variables – recidivism and company size. Overall, we found that new entrants outperformed incumbents, and small companies had an advantage over larger companies. We will consider our findings relative to these variables in turn. The findings have complementary ties to existing literature and merit either further experimentation, policy change, or some combination thereof.

Our initial hypothesis was that small business commercialization rates would improve as firms gain experience working with the government. Our results did not support that hypothesis; rather, they revealed a more nuanced behavior. In considering just the upper and lower quartiles of our set, new entrants to SBIR (fewer than 4 contracts, median 3) had statistically indistinguishable performance from incumbents (36–419 contracts, median 102). We hypothesized that commercialization performance would improve as companies gained experience; however, the performance of the most experienced companies (median of 102 contracts) was indistinguishable from those with the least experience (median of 3 contracts). In expanding our analysis to compare the upper and lower halves of our population, we found that our initial hypothesis is reversed: the new entrants (fewer than 14 contracts) outperform the incumbents (15 to 419 contracts) by 2.6%. This difference in performance is driven entirely by companies in the second quartile (5 to 14 contracts). These companies outperform all other quartiles by 6%. If we limit our hypothesis test to a comparison between the first quartile (1 to 4 contracts) and second (5 to 14 contracts), we see evidence of a learning effect – a 6% increase in performance. However, commercialization performance drops by 5.9% in the next quartile and stays at that level into the fourth quartile. This spike in performance in the second quartile warrants further consideration. While learning may explain the increased performance witnessed in the transition from the first to the second quartile, what dynamics are driving the 5.9% drop in performance in companies with more than 15 contracts?

This spike in performance in the second quartile possibly represents a convolution of effects: on one side, the expected benefits from learning; on the other, a separate dynamic. The literature provides a possible explanation for the subsequent drop in performance, the "SBIR mill" phenomenon (Lerner, 2000; Link and Scott, 2009). SBIR mills exploit the public policy underlying the SBIR program. They are firms that exist, at least in part, for the purpose of securing SBIR awards with no intent to commercialize (rent-seeking). These firms may be less innovative and less likely to commercialize than other firms (Link and Scott, 2009). They are alternately known as "frequent winners," a class of firms that underperforms yet accounts for a disproportionate rate of awards (Federal Research, 1999). The rent-seeking behavior of SBIR mills would account for a class of firms with high recidivism and low performance – a dynamic that may underlie the drop in performance beyond the second quartile.

The goal of the SBIR program is to encourage high-tech innovation in the USA. The DoD invests in those areas of interest to national defense. The scope of this article is limited to 433 Air Force SBIR projects from 2015 to 2018; other Services and organizations may incentivize adoption differently. Analysis indicates that the average SBIR company in this data set had 39 contracts. This represents $6.7 million to $39 million in SBIR funding and 19.5–78 years in periods of performance. If no benefit is derived from recidivism, or worse, if firms have less than earnest intent, a limit to recidivism should be considered. Reducing recidivism or setting limits on recidivism is in line with the intent of the SBIR program. Awards of over 100 contracts (or more than 400) to a single firm provide repeated stimulus for a single firm versus an industrial base. Since the objective of the SBIR program is the stimulation of an economic base, a policy that limits participation to companies with fewer than 100 historic government contracts (or perhaps fewer) merits consideration. Our first independent variable was recidivism – our second was firm size. Consistently with the innovation literature, we found that smaller firms outperformed larger firms. Further, and perhaps a confounding of positive effects, new entrants (low recidivism) were on average, smaller companies.

Businesses with fewer than 31 employees (the lower half of our data) have a commercialization rate that is 2.6% higher than larger businesses (32 to 499 employees). Further experimentation and research are merited to determine effects based on company size; the overall low rate of success limited our ability to draw statistically significant conclusions for finer gradations (i.e., lower than quartiles) based on company size.

In considering our two independent variables, we find improved performance among new entrants (1 to 14 contracts) and smaller firms (1 to 30 employees). The improvement from favoring either new entrants or smaller firms is 2.6%. On the surface, this improvement appears modest. However, the global commercialization performance of this set was 8.8%; relative to that baseline, 2.6% is significant. To give some scale to this number, this data set represented 433 individual capability-development efforts; 2.6% translates to 11 novel capabilities. Further, considering the DoD investments ($1.8 billion in FY 2019), 2.6% represents roughly $47 million in investment. Small percentage gains from policy shifts can have real effects on the development of needed capabilities and on the efficacy of our investments.

RECOMMENDATIONS

Our analysis considered commercialization performance within Air Force SBIR investments. An external calibration or comparison to other similar programs is warranted. Comparison of commercialization rates of Air Force SBIR programs should be made to other SBIR programs as well as innovation programs in the commercial market. The approximately 8.8% rate of commercialization for Air Force SBIR programs may or may not be comparable to the innovation rates of the broader market. Such a comparison would help determine the relative success of the SBIR program and other opportunities to improve. This would inform value provided to the Warfighter. This future research could be furthered by comparing areas of innovation of SBIR programs with similar commercial endeavors.

Further research should be accomplished, specifically experimentation with policy to deliberately target new entrants or limit the number of previous awards allowed. This form of policy experimentation is in line with the objectives of the SBIR program. SBIR works to stimulate high-tech innovation in the USA while targeting specific research and development needs of the government versus enriching individual firms (SBIR | STTR, n.d.). The program does not meet its goal by repeatedly funding the same small businesses with no increased commercialization rate. Program eligibility and selection criteria could consider the number of previous SBIR awards. An investigation into the bump in commercialization rates in the second quartile would complement this research as well. A better understanding of the key dynamics that result in both the increase and subsequent decrease in performance could better inform limits to recidivism as well as other policy choices. This could be explored through analysis of learning effect for businesses.

Experimentation could also take place to further evaluate the performance of "small." Again, limitation through program eligibility requirements or evaluation criteria of select programs could assist in confirming findings. Additionally, research should refine "small" company size. More gradation between a small company of 1 employee and a small company of 500 employees is needed. With a larger data set, those break points could be determined. Further research should take place to determine the value associated with the Air Force SBIR programs. This should be conducted in two parts:

- First, by examining the overall value for money of the approximately 8.8% of programs that were commercialized. The commercialized Air Force SBIR programs could possibly represent a substantial return on investment.
- Second, by examining the value to the Warfighter. Advancements in capability or value provided to the Warfighter may be procured that are not represented strictly by commercialization.

REFERENCES

Badiru, Adedeji B. (2020). *Innovation: A Systems Approach*, Taylor and Francis/CRC Press, Boca Raton, FL.

Badiru, Adedeji B. and Cassie B. Barlow, Eds. (2019). Defense Innovation Handbook: Guidelines, *Strategies, and Techniques*, Taylor and Francis/CRC Press, Boca Raton, FL.

Badiru, Adedeji B. and Gary Lamont (2022). *Innovation Fundamentals: Quantitative and Qualitative Techniques*, Taylor and Francis/CRC Press, Boca Raton, FL.

Balzat, M. (2006). *An Economic Analysis of Innovation: Extending the Concept of National Innovation Systems*, Edward Elgar Publishing, New York.

Blake, E. (2020). *Determinants of Small Business Innovation Research Performance* [Master's thesis, Air Force Institute of Technology]. AFIT Scholar. https://scholar. aft .edu/etd/3227/

Bresler, A. (2018, May 9–10). Bridging the Gap: Improving DoD-Backed Innovation Programs to Enhance the Adoption of Innovative Technology Throughout the Armed Services [Paper presentation]. Acquisition Research Symposium 2018, Monterey, CA. https://apps.dtic.mil/sti/pdfs/AD1069599.pdf

Fagerberg, J. and D. C. Mowery, Eds. (2006). *The Oxford Handbook of Innovation*. Oxford University Press. https://doi.org/10.1093/oxfordhb/9780199286805001.0001

Federal Research: Evaluation of Small Business Innovation Research Can Be Strengthened, Subcommittee on Technology, Committee on Science, House of Representatives, 106th Cong. (1999). (statement of Susan D. Kladiva). https:// www.gao.gov/assets/t-rced-99 -198.pdf

Joint Chiefs of Staff. (2018). *Joint Capability Areas: Force Structure, Resources and Assessment (J-8) Directorate.* https://intellipedia.intelink.gov/wiki/Joint_ Capability_Areas

Kelly, J. and R. Sensenig (2019). Department of Defense: SBIR/STTR Grants and Other Contracts. *Academic Entrepreneurship for Medical and Health Scientists*, Vol. 1, No. 3. https://repository.upenn.edu/ace/vol1/iss3/14

Kirsner, S. (2018). The Biggest Obstacles to Innovation in Large Companies. *Harvard Business Review.* https://hbr.org/2018/07/the-biggest-obstacles-to -innovation- in-large-companies

Lerner, J. (2000). The Problematic Venture Capitalist. *Science*, Vol. 287, No. 5455, pp. 977– 979. https://doi.org/10.1126/science.287.5455.977

Link, A. N. and J. T. Scott (2009). Private Investor Participation and Commercialization Rates for Government-Sponsored Research and Development: Would a Prediction Market Improve the Performance of the SBIR Programme? *Economica*, Vol. 76, No. 302, pp. 264–281. https://doi.org/10.1111/j.1468-0335.2008.00740.x

Mattis, J. (2018). Summary of the 2018 National Defense Strategy. Office of the Secretary of Defense. https://dod.defense.gov/Portals/1/Documents/pubs/2018- National-Defense -Strategy-Summary.pdf

Quinn, J. B. (1985). Managing Innovation: Controlled Chaos. *Harvard Business Review*, Vol. 63, No. 3, pp. 73–84. https://hbr.org/1985/05/managing-innovation-controlled-chaos

Rask, T. (2019). *Commercialization Analysis of SBIR Funded Technologies* [Master's thesis, Air Force Institute of Technology]. https://scholar.aft.edu/etd/2351/

Ryan, K., A. Cox, E. Blake, C. Koschnick, and A. E. Thal (2022), Innovation Transition Success: Practice Doesn't Make Perfect. *Defense ARJ*, Vol. 29, No. 4, pp. 336–358. https://doi.org/10.22594/dau.21 872.29.04. Note: Adapted and reprinted with permis-sion from Defense Acquisition Research Journal.

SBIR I STTR. (n.d.). *About the SBIR and STTR Programs.* Small Business Administration. Retrieved June 22, 2020, from https://www.sbir.gov/about

Schilling, R., T. A. Mazzuchi, and S. Sarkani (2017). Survey of Small Business Barriers to Department of Defense Contracts. *Defense Acquisition Research Journal*, Vol. 24, No. 1, pp. 2–29. https://doi.org/10.22594/dau.16-752.24.01

Serenko, A., N. Bontis, and T. Hardie (2007). Organizational Size and Knowledge Fow: A Proposed Theoretical Link. *Journal of Intellectual Capital*, Vol. 8, No. 4, pp. 610–627. https:// doi.org/10.1108/14691930710830783

Small Business Administration. (n.d.). Birth and History of the SBIR Program. Retrieved October 2, 2020, and August 28, 2021, from https://www.sbir.gov/birth -and- history-of-the-sbir-program

Small Business Innovation Development Act of 1982, Pub. L. 97–219. (1982). https://www .congress.gov/bill/97th-congress/senate-bill/881

Van de Ven, A. H. and M. S. Poole (1989). Methods for Studying Innovation Processes. In H. Van de Ven, H. L. Angle, and M. S. Poole, Eds. *Research on the Management of Innovation: The Minnesota Studies*, Harper and Row, New York, pp. 31–54.

Index

Printed in the United States
by Baker & Taylor Publisher Services